To Connect is to Understand Mathematics 2

Selected works 2007-2014

ALFINIO FLORES

DEDICATION

To the loving memory of my parents Margarita and Humberto, who prepared me to walk around the world and encouraged me to find my way.

CONTENTS

ACKNOWLEDGMENTS

The articles were published before in the journals, proceedings, and books indicated. The complete reference to the original source is given in a footnote in the corresponding page. I appreciate the free and prompt permissions received to republish these articles from these publishers:

PRIMUS (Taylor and Francis)

Ohio Journal of School Mathematics (Ohio Council of Teachers of Mathematics)

The Centroid (North Carolina Council of Teachers of Mathematics)

International Journal of Mathematical Education in Science and Technology (Taylor and Francis)

Journal of Mathematics Education at Teachers College (Teachers College, Columbia University)

Mathematics Enthusiast Journal (Department of Mathematical Sciences, University of Montana)

Teaching Mathematics and its Applications (Oxford University Press)

Journal of Mathematical Behavior (Elsevier)

Proceedings Third and Fourth Conferences on Mathematics Education Research in Texas (Mathematics Education Research in Texas MERiT)

Cálculo diferencial e integral para profesores, formadores e investigadores en matemática educativa. Estudios y reflexiones sobre su enseñanza (Pearson)

Promoting high participation and success in mathematics by Hispanic students: Examining opportunities and probing promising practices (Todos: Mathematics for all)

Thanks to the direct intervention of NCTM's president Matt Larson I was granted permission by NCTM in 2016 to include in this collection my articles from the journals *Mathematics Teaching in the Middle School, Mathematics Teacher,* and the books *Understanding geometry for a changing world,* and *Empowering the mentor of the experienced mathematics teacher.*

I want to thank my coauthors of articles in this volume: Melina Priewe, Jaclyn Braker, Kelly Kimpton, Kevin Cauto, Alexandrea Hammons, John Pelesko, Chuck Biehl, Mary Ann Huntley, Irene Gallegos, Kay Biondi, Cheryl Thomas, Emily Gustafson, H. Bahadir Yanik, and Jeong Oak Yun. It was a pleasure to work with them. Their names appear in the reference to the published version.

The version of the articles included in this collection stems from my own files. In some cases they do not include the last editorial improvements from the publishers.

1 OPENING THE DOORS OF OPPORTUNITY

Introduction

This article highlights some of the ways mathematics teachers can help open the doors of opportunity for their students, especially second language learners. It is based on examples from observations of the teaching of Ms. Adelaide[1], and from conversations about teaching practices in person and via email. Ms. Adelaide's district began offering content classes in English as a Second Language (ESL) and Bilingual formats in the fall of 1991. By the mid 1990s, her state's Department of Education created K – 12 standards along with a high school exit exam. During that time, she completed her Master's Degree; for her thesis, she researched effective teaching practices and developed her own ideas for her math ESL and bilingual classes. Her state has recently joined other states in implementing Common Core State Standards, and although she is now retired, many of her ideas are still pertinent. Shulman has highlighted the importance of the wisdom of practice in education and pointed to the necessity "to study accomplished practice as it actually occurs and to ask how it has been achieved." (Shulman, 2004 p. 252)

Of course there are many other teachers who find ways of making learning mathematics easier and interesting while instilling confidence in students when it comes to trying something new, but in the case of Ms. Adelaide's practices they illustrate the importance for teachers to know the content subject well, know their students well, and to present the mathematics artistically.

Ms. Adelaide taught mathematics at Votech High School[1], an urban magnet school in a large metropolitan area in the Southwest. At the time Ms. Adelaide taught there, the school offered both academic and vocational programs that allowed students to have a choice for their futures. The school encouraged students to go to college, while at the same time offered 30 career-vocational programs such as Electrical Wiring, and Welding.

---

[1] name is a pseudonym

Counselors actively promoted programs, like *Adelante* and *Access to College Education*, that built connections to school and developed students' expectations to continue through college.

Her students were mainly from groups that traditionally have not been served well by state, district or school policies, or by the way mathematics courses are traditionally organized and taught in schools. Students at Votech were generally middle or low socio-economic status and many faced difficult obstacles. Some students were brought as children into the country without documents and faced an uncertain future as most jobs require documents. These students know they will have to create their own future and will have to make it on their own, as economic support from the family may or may not be there. They work long hours.

Some of her students were recent immigrants and faced the daunting task of learning a new language at the same time that they needed to learn new content. Learning a new language as a teenager is not easy. Teenagers are sensitive and think that people will laugh at them. Some students come from families where English is not the dominant language and have poor English skills even though they may have been born and raised in the U.S. Their language problems included difficulties with noun agreement, verb conjugation, mixing up past and past perfect tenses. They also had problems using the right expressions to describe mathematical steps and operations. Unhappily, there were also some smart-aleck friends who told them things incorrectly just for laughs.

A flexible math teacher

Students knew Ms. Adelaide was an exacting and strict teacher. If seniors in her class had an F, she made it clear they were not going to graduate if they were failing her class. One student was in the habit of getting his own way, and would not do the work. He did not believe Ms. Adelaide would actually flunk him. However, after he saw she was serious, he ended up being a "true believer" and did the work. Former students, many of whom complained and whined about the amount of work in her class, often came back to visit her after they realized how much they learned in her class, or when they realized they were not learning nearly as much from later teachers. Many former students came to her with a variation of "I hated you when I was in your class, but I'm glad I was."

A native English speaker who had failed the state test twice finally agreed to enroll in Ms. Adelaide's class as the girl's counselor had suggested. Ms. Adelaide carefully checked work for the homework, not just the answers, and realized the type of mistakes the girl made. Ms. Adelaide showed the girl where she got mixed up and allowed her the chance to fix the work for a better grade. The girl passed the state test the next time she took it. Ms. Adelaide often reminded the students that THEY passed the test. She merely helped them prepare for the test.

One way in which Ms. Adelaide showed flexibility was in accepting, and even encouraging, alternative methods, algorithms, or strategies from students, whether invented by them, learned from other teachers (Perkins & Flores, 2002b), or in other countries. Although she may not have known all the ways in advance, she was always willing to try to understand why the alternative ways worked. The teacher became the learner and in addition she would build learning opportunities for all students based on an alternative method.

Ms. Adelaide assigned homework and expected all students to complete the work. Ms. Adelaide knew which of her students had to work. In such cases, she winked at their late work. Some students were young mothers. One was 17 with two children. Ms. Adelaide quietly accepted her work when she got it in. The other students in her group did not say a word about this "freebie".

She was also flexible when students made silly mistakes. One student did the homework thoroughly but used the wrong formula for an entire section of the homework. She showed him his mistake and allowed him to fix the work for a better grade. She also allowed all students to rework their homework errors for a better grade. They would staple a new sheet of paper to the original homework and write "Re-Read" and re-submit the entire packet for a better grade. Some students took advantage of the re-read; some did not.

Although students knew they could have extra time when they needed it, they also knew the work still had to be done. Ms. Adelaide told students she was not a "Nice guy." As her students would soon find out, the world is not peopled by those who give second chances. If they don't make the car payment, they lose the car. If they don't pay the electric company, there are no lights or air conditioning.

Ms. Adelaide made herself available beyond the school hours. She arrived at school early, at 7:00 am, and the doors of her classroom were open for students at 7:30 even though the first period did not start until 8:25. The doors of her room were open for students who needed extra help also after school and during lunch, as students knew she brought her own lunch. Students were welcome to have their lunch there too (there were paper towels and 409 available; no student ever left a mess for others to pick up).

Coping with adverse policies and institutional barriers

State mandated testing and limits on teachers use of the students' native language impacted teaching approaches. However, she used every talent she possessed. If a student expressed himself in Spanish, she didn't stop him but made sure he learned the English version of the concept. Ms. Adelaide did her instruction in English, but at the beginning of the year she encouraged students to speak to her either in English or Spanish. She would answer questions in either English or Spanish; her choice of language depended on who asked the question and the level of difficulty of the response.

Her state uses local property taxes as the main way to fund schools. This results in heavy inequities in funding for poorer districts like hers. Ms. Adelaide also dealt with district and school policies and regulations that had an effect on her teaching. Some administrative choices sometimes resulted in not buying enough books to have one for every student. The district wanted students to work in groups of four. The district had assumed that students would work cooperatively if they had to share textbooks. Consequently, Ms. Adelaide had 16 books instead of 32. Students and teacher were not happy about the situation.

Although big tables appropriate to work in groups and build geometrical objects were replaced by small individual tables by the administration, Ms. Adelaide's room was arranged so that students could work individually and in groups of four. Seats were placed so that all students could see the white board, the screen, and the teacher up front. There was a computer with a ceiling projection system. The walls of Ms. Adelaide's classroom were covered with formulas and figures related to the mathematics she taught, including commercially available or teacher made posters, and student work. The assistant principal described her classroom as a walk-in textbook. All teachers were required to cover all the posters in their rooms before the state mandated test.

The school sometimes made enrollment decisions that were detrimental for her students. One year, her geometry class for second language learners was smaller than when they were in algebra, because the top performing students were placed in mainstream geometry honor courses. Ms. Adelaide objected to that because the language used in geometry was quite more complex than everyday English. Her top students, although quite capable in terms of the mathematical thinking, still required the support with language that was not available in the mainstream course. The students who were removed from her class did survive one week in the honors geometry course. The students requested to be put back with Ms. Adelaide. Three of those students were in period 1 class and were the top three out of 24, and the other student was #2 out of 25 in period 8. However, the other students in the second language learners section, without the top third in their classroom, did not perform as well nor were they as motivated as the previous year when all were together. The negative impact of tracking on the middle and lower performing students has been documented by research (Oakes, 1990).

On the other hand, the equity and fairness battles Ms. Adelaide had helped gain over the years contributed to her present students having the same opportunities to learn quality mathematics as other students. She made sure English language learners would have access to the same content in mathematics and use the same textbooks. No longer were second language learners students given textbooks written at the elementary school level.

Incorporate students' native language and background as resources

In some cases, students were schooled in other countries where the sequence in mathematical topics was different. Some students came with knowledge of the content of the first semester of the new school's algebra and the first semester of geometry courses, but were lacking the content of the second semester of each course. The mismatch created problems for the students. Ms. Adelaide researched textbooks and the sequencing of the topics in other countries. Her findings gave her insight to the frustrations both teachers and students felt and she shared her findings with other teachers. Because the first semester was more like a review for these students, it gave her time to build their English mathematics vocabulary. Ms. Adelaide learned about the notation and procedures her students learned in their former countries (see, for example, Perkins and Flores, 2002a), and she used this knowledge to the advantage of her students.

Ms. Adelaide took special care to enunciate words carefully and avoided the use of slang. She also addressed issues of grammar, pronunciation, and spelling as she discussed mathematical ideas with her students. Because there is not always a close and regular correspondence between the way words are pronounced and written in English, this constituted an additional hurdle for her students. Ms. Adelaide took the time to emphasize the different sounds of the letter *c* in *isosceles* and in *scalene*, and compared this variation to the pronunciation of *c* in Spanish. Frequently she wrote on the board what she and the students said so that students had also a visual support to follow the conversation. Also, students would hear slang on campus, come in and write it phonetically on the board, and asked "¿Qué es esto?" It was a teaching moment. They trusted her to explain English phrases to them. To them it was just as important as last night's homework.

The students would literally stop the class when Ms. Adelaide used slang occasionally or an idiomatic phrase in class. To her, it was part of their integration into American English. It was not "lost time" in mathematics.

Her state passed some years ago a law prohibiting teachers the use of Spanish and other languages for instruction. Ms. Adelaide observed the law; she no longer used Spanish for instruction. However, allowing the students to ask questions in Spanish was a form to validate their knowledge and value the use of Spanish in their learning. Ms. Adelaide made it very clear that it was quite appropriate for students to use whatever language they felt more comfortable to share their thinking. Ms. Adelaide treated the language of her students as a legitimate language for academic work. Some students spoke mainly in English, some mainly in Spanish, and some used code switching. Here is one example.

Student: Se supone que tiene que dar dos medidas iguales.

Teacher: Two measures are equal.

Student: Creo que es un right triangle

In terms of English usage, Ms. Adelaide did not correct what students said. She simply rephrased what they said using academic English and mathematical precise vocabulary. She knew students would not learn academic English or mathematics if its use was not modeled. When students read the mathematical notation incorrectly, in a smooth way she immediately used the correct terms. This example aroused with $x\sqrt{2}$

Student: x to the square root of 2

Teacher: x times the square root of 2

Students discussed in small groups, shared ideas, and then presented the ideas of the group to the whole group. Sometimes a consensus was reached in the small group and sometimes students shared different opinions. Ms. Adelaide made sure all students had a chance to voice their ideas.

Teacher: Alejandra did not have a chance to say anything.

Establishing personal connections with students

Ms. Adelaide also took the time to establish personal connections with her students. She used humor in her teaching and often let students have a good chuckle at the expense of the teacher. She shared stories about her family and favorite moments and pets.

Many of her students had to work to put food on the family table. Often, students believed she did not have to struggle in life like them, that she was rich and could afford everything she wanted, whether a new car or a trip. She explained that everything she had, she worked for it, and so could they and they could have those things, too. She often told stories of her childhood and family, of growing up in hardship, of overcoming great obstacles, and suffering great losses in life.

Her openness to discuss death of loved ones paid great dividends. In each year, there were always a few deaths among both students and staff. Students felt they could talk to her about it. Ms. Adelaide's advice to teachers is to not ignore death. If teachers are uncomfortable, they can arrange for one of the counselors to meet with the student even if it's during their own class time. The important person in this instance is the grieving student.

Her last two assistant principals for curriculum instituted student use of binders. As the students took the chapter test, the teacher evaluated their binders. She set up her own grading standards; it was an easy 10 points for keeping a neat and complete binder. Ms. Adelaide created her own dividers to separate students' work into weekly grade reports, homework, study guides, class notes, and Questions of the day along with the grading page for the binder. Each separator had the rules for each topic.

The binder's grading page provided a place for teacher to student communication, both formally and informally. And, the students being teenagers decorated the binders as they saw fit. As students were free to embellish their binders, Ms. Adelaide learned a great deal about her students by reviewing the binders.

Students often kept memorial cards in the binders. One tough guy had the speech he read at his friend's funeral.

> Teacher: "Did you write this?"
>
> Student: "Yeah."
>
> Teacher: "It is very good."
>
> Student: "Thanks."

Student and teacher got along better after she read that eulogy.

One student had a photo of her little sister all dressed up. Ms. Adelaide commented on how nice she looked and if it was a special occasion. The student explained the photo was taken on the day her sister was presented in church at age 3 in thanksgiving of her life, and explained more about the ceremony. The students liked that Ms. Adelaide thought it was very nice.

In turn, students learned about Ms. Adelaide from her reactions to the binders. One girl had a poem in which a child who has died is talking to her mother. The child tells the mother that she was chosen to be the mom because she would be the perfect mom for this child who would have a very short time on Earth. Ms. Adelaide wrote a note asking the student for a copy of the poem; she told the student it spoke to her. She also told the girl about her own daughter, her asthma and that she had died at 23.

Binders helped in communications with parents. At conferences, parents and teacher would go through the binder page by page together. Parents could ask questions and she would answer. By conference's end, Parents had a better understanding of class expectations and of their own student's class performance.

High-stakes tests

Ms. Adelaide included daily practice for the high-stakes exit exam. If students did not pass the exam, they would not graduate from high school and many doors would be closed for them. The daily warm-up was based on the state test's formula page and problems the class may have stumbled over in homework.

Ms. Adelaide reviewed the materials provided by the state to determine what specific topics were covered in the state test, and include those in her teaching.

She also gave instructions to her students on how to take the test. Several students told Ms. Adelaide they heard her voice during the test. "Do all the steps! Take your time. Where is the formula page?"

Use of concrete and visual teaching tools

While the state exit test was important, Ms. Adelaide also wanted her students to think about and find various approaches to problems. Ms. Adelaide used many hands-on activities and materials to make mathematics relevant for her students. For example, students used their own hand-made clinometers to measure the height of a palm tree. The clinometer was made of a paper protractor with a rotating soda straw. Ms. Adelaide had photographs of clinometers in a crane, in a boat rudder, and in an airplane. She discussed how the crane operator needs to pay attention to the angle of the arm of the crane. When the angle is too low, the crane could flip over.

Ms. Adelaide knew well the mathematics she taught. She established connections among topics, and illustrated concepts with real life examples. She also had a good understanding of what makes ideas difficult for students, what are the most common misconceptions students have, and what are the examples that work particularly well to help students understand a specific concept (Shulman, 1986). For instance, when giving the relations between the legs and hypotenuse of the special triangle 45-45-90 the book and formula sheet pictured the triangle on one of its legs. Ms. Adelaide brought to the attention of students cases where the triangle was on the hypotenuse and how this triangle was simply rotated and that the relations between the sides were still the same. Ms. Adelaide included also examples where students had to reverse the process. There were examples where the hypotenuse was given, so instead of multiplying the length of one leg by $\sqrt{2}$ to obtain the hypotenuse of an isosceles right triangle, they had to divide the length of the hypotenuse by $\sqrt{2}$ to obtain the leg of the triangle.

Before looking formally at the next special triangle, Ms. Adelaide handed out a set of plastic triangles to each of her students. One triangle was familiar to them (45° angle). She asked them to estimate the angles of the other triangle. Students showed good sense of the sizes of angles in their estimations. She asked how the two triangles were different and how they were the same. As students overlapped and compared the triangles and discussed with each other, they had opportunity to get acquainted with the shapes and make discoveries on their own. Two 45-45-90 triangles together can form a square when sharing a hypotenuse, or a bigger 45-45-90 triangle when sharing a leg. Two 30-60-90 triangles can form an equilateral triangle.

In another lesson, as students worked with the formulas for the volumes of pyramids and cones, Ms. Adelaide held a student-made pyramid or a cone in her hand and with a stick illustrated what was the height used to compute the volume. She opened the pyramid so students could see inside, and placed the stick to go from the apex of the pyramid to the center of the base, so that the stick was perpendicular to the base. Then she placed the stick along one of the triangular faces to show the slanted height used to compute the surface area. She also illustrated, by placing the cone top down, how some "other" students thought that a different formula may be required to compute the volume of a cone, because the base was no longer at the bottom. She told about how a younger child thought that the shape with the vertex pointing down was no longer a cone, but was "like what they use to hold ice cream."

Ms. Adelaide understood many of her students would be parents soon; some already were. She wanted them to understand toys and everyday objects could be used to teach their children. Froggie, duck, puppy and

Garfield were there to help her plant the idea of using toys to teach. The faces of her students lighted up when she included the toys.

In algebra, students enjoyed playing the shark and boat game to add and subtract negative and positive numbers (Lovitt and Clarke, 1988). They also enjoyed playing battleship, including the girls. Ms. Adelaide also had a set of markers that snap together on her desk. Student would stand there talking to her and they would be constructing something. She did not take their creations apart but left them on the desk for others to see and create something else.

Ms. Adelaide used many different kinds of tools for learning and let students explore on their own with them. In an exploratory activity, all Ms. Adelaide did was to show how to join the Polydron pieces, and students built their own shapes: icosahedron, a non-convex deltahedron, one quarter of a cylinder, pyramids (square, triangular, pentagonal), half a sphere together with a cone to serve as a balancing toy, a "soccer ball" (a truncated icosahedron, formed by 20 regular hexagons and 12 regular pentagons), and other complex shapes formed by joining the shapes constructed by individual students.

On a day she would be absent, she left a video on the platonic solids, a work page, and an invitation to make icosahedral maps of the Earth. Step structures made of paper prisms were on display. They were part of an exploratory activity she used on the first day she showed them nets. An empty box of detergent and a bunch of scoops were in the back of the room. As extra credit students could figure out a way to determine the volume of the scoop, which resembled a somewhat rounded truncated pyramid.

Ms. Adelaide had a collection of small stuffed toy animals that showed symmetry: a frog, a duck, a puppy, etc. She asked her students: If this were a real frog, and you cut it open, would it look symmetric in the inside? She would also ask if any students were taking biology and if so, had they covered linear symmetry in that class. She also had non-examples: Garfield sitting sideways was not symmetric. There was also much discussion about the animal whose label made him asymmetrical; without the label he would be symmetrical.

In some of her favorite lessons, food was used as a tool for teaching. For example, when students learned about measurement they made ice cream with a tin can (Flores & Perkins, 1996). In another activity students used popcorn to learn about volume and develop a sense for it. Students learned about ratio, too: "If one measure of popcorn can make enough to serve 8 kids, how many times will I have to run the popcorn popper to have enough for everyone?" Students also used lemonade to develop the concept of ratio. These exercises were discussed with principals and department heads ahead of time so no one would think they were merely class parties.

Final comments

Sometimes, when people saw Ms. Adelaide teaching in her classroom and observed how she met the needs of her students who were learning English as their second language, they may have thought it was just good teaching, that she did not do anything special. Indeed, what Ms. Adelaide did would constitute good teaching with any other group. Other students would clearly also benefit from her approach to teach mathematics.

However, what is interesting about Ms. Adelaide's teaching is that she made sure to incorporate elements of good teaching that are especially relevant for second language learners. In other words, she made sure to incorporate techniques into her teaching whose omission is especially detrimental to students who are struggling with their new language. Furthermore, Ms. Adelaide did address the needs of her students in terms of language development, cultural differences, and other aspects that are crucial for English language learners (Bay-Williams & Herrera, 2007). In her classroom students developed English and mathematics proficiency, which would open many doors of opportunity for them. At the same time, they developed love for learning, which is Ms. Adelaide's great legacy.

References

Bay-Williams, J. M. & Herrera, S. (2007). Is "just good teaching" enough to support English Language Learners?: Insights from Sociocultural Learning Theory. In W. G. Martin and M. E. Strutchens (Eds.), *The Learning of Mathematics* (pp. 43-63). Reston, VA: NCTM, 2007.

Flores, A. & Perkins, I. (1996). Tin-can ice cream. *School Science and Mathematics*, 96, 46-49.

Lovitt, C. & Clarke, D. (1988). *The Mathematics Curriculum and Teaching Program. Activity Bank* (Vol. 1). Canberra, Australia: Curriculum Development Centre.

Oakes, J. (1990). *Multiplying Inequalities*. Santa Monica, CA: RAND Corporation.

Perkins, I. & Flores, A. (2002a). Mathematical notations and procedures of recent immigrant students. *Mathematics Teaching in the Middle School,* 7, 346-51.

Perkins, I. & Flores, A. (2002b). Why don't teachers know all the ways? *Mathematics Teaching in the Middle School,* 7, 262-263.

Shulman, L. S. (1986). Those who understand: Knowledge growth in teaching. *Educational Researcher, 15*(2), 4-14.

Shulman, L. S. (2004). *The wisdom of practice.* San Francisco, CA: Jossey-Bass.

2 HELPING PROSPECTIVE TEACHERS OF CALCULUS FOCUS ON CONCEPTUAL UNDERSTANDING[2]

Abstract

This chapter addresses ways in which we can help future teachers of high school develop a better conceptual understanding of calculus for teaching. The majority of the activities and ideas presented are part of a methods course that includes activities, readings, and discussions geared to give future teachers a better conceptual understanding of calculus, as well as get them acquainted with some of the issues students face related to the learning of calculus. First we present activities related to the limit of a sequence using graphing calculators, pictorial representations, paper strips, and a dynamic geometry program. Next we address topics related to the derivative, such as the method of increments, the slope of secant lines, relate the area of a circle to its circumference, magnify graphs of functions to see which ones look locally straight, and discuss linear approximations to a function and the corresponding error. Then we use a computer program to graph approximations to the derivative function using quotient of increments, and discuss how motion detectors can be used with a bouncing ball to study the relation of a function to its derivative function. Next, concepts related to the definite integral are introduced using motion, area, and Riemann sums. Logarithms are used as an example. The next activities

---

[2] Flores Peñafiel, A. (2014). Ayudando a futuros profesores a mejorar la comprensión conceptual del cálculo. In Pluvinage, F & Cuevas, A. (Eds.), *Cálculo diferencial e integral para profesores, formadores e investigadores en matemática educativa. Estudios y reflexiones sobre su enseñanza.* México City: Pearson. English version printed by permission.

address the indefinite integral and the fundamental theorem of calculus. At the end we discuss how teachers can identify and address some common misconceptions of students about limits, and get acquainted with stages in the understanding of the concepts of derivative.

Computational proficiency in calculus and conceptual understanding

Calculus, as its name implies, is an impressive and powerful tool for calculation. At the University of Delaware, like in many other universities in the United States and other countries, most students who want to become secondary mathematics teachers are highly proficient in the techniques of calculus. Some of them start to develop this proficiency already in high school. They take the Advanced Placement course for calculus and receive college credit for calculus. The Advanced Placement exam in calculus mainly measures proficiency in computation and the ability to apply the techniques, rather than conceptual understanding (for sample exam questions, see College Board 2011). It is therefore not surprising that high school students focus almost exclusively on procedures rather than developing also a conceptual understanding. When students arrive at college, the Advanced Placement credit allows them to skip the first calculus course. One might think that they miss an opportunity to have a more conceptually oriented calculus course in college. Unfortunately, in many places, at the college level, courses in calculus still emphasize procedural proficiency rather than conceptual understanding. Therefore, many future teachers may never have the opportunity to experience themselves a conceptually based calculus course. As teachers, they will in turn very likely continue propagating the cycle and emphasize mostly procedural fluency.

However, the calculus is also a rich body of interconnected concepts and relations. Prospective teachers need to have the opportunity to develop a rich understanding of each of the main concepts in calculus, and explore in depth the relations among the concepts. For example, in relation to the derivative of a function f at a point $x = x_0$, there are several conceptions associated (Artigue 1991, p. 175). We can view the derivative as the limit of the ratio $\dfrac{f(x_0 + h) - f(x_0)}{h}$ when h tends to 0. Or we can think about the best linear approximation to the function f at x_0, either as the coefficient characterizing the linear map tangent to f at x_0, or as the coefficient of the first order term of the series expansion of f around x_0, either for the full series (Lagrange) or of the expansion limited to order 1. Or we can think of the slope of the tangent line at x_0, or as the slope of a highly magnified portion of the graph that looks "locally straight". In contrast to these more conceptual approaches, many students only see derivative as the number obtained by applying the usual rules of differentiation, using the derivatives of the elementary functions.

Each of the conceptual approaches mentioned above in turn can be unpacked to make more explicit the different concepts involved. For example, before students understand the limit $\lim_{h \to 0} \frac{f(x_0 + h) - f(x_0)}{h}$, they need to understand the quotient of increments $\frac{f(x_0 + h) - f(x_0)}{h}$, and need to establish connections between the quotient of increments with other concepts such as the slope of a secant line to a curve, or the average speed in an interval for an object moving at variable speed. Then, after they have a good grasp of the ratio of increments, students need to see how the limit of these ratios is related to the slope of a tangent line, and to the instantaneous speed of the moving object at a given time. Afterwards, they need to establish the connection between the limit at a given point, and the derivative function. Although the notation for the derivative at a point, $\lim_{h \to 0} \frac{f(x_0 + h) - f(x_0)}{h}$ is very similar to that of the derivative function $f'(x) = \lim_{h \to 0} \frac{f(x + h) - f(x)}{h}$, they are quite different concepts. One is a number and the other a function. Conceptually it represents a huge difference, although sometimes textbooks introduce the definitions for both the derivative at a point and the derivative function in the same paragraph.

Thurston (1998) includes vision, spatial sense, kinesthetic (motion) sense as one of the major divisions that are important for mathematical thinking. In some cases, students will figure by themselves meanings behind the mathematical symbols, and talented students will use kinesthetic or visual approaches to develop a deeper understanding on their own. For example, when my son Luis was taking calculus in High School he asked me why we did not use scissors to explain ideas in calculus. I asked him what he meant and he described how when we cut a curve with scissors, the direction of the scissors in the moment of cutting would be the direction of the tangent line (Figure 1). He also pointed out that scissors rotate as we cut along the curve, and that the rotation could be related to concavity and points of inflection. For example, when scissors change the direction of rotation, that would be the point of inflection (Figure 2). (For more examples, see Flores and Flores 2005, 2006).

Figure 1. Direction of cut and tangent line to a curve.

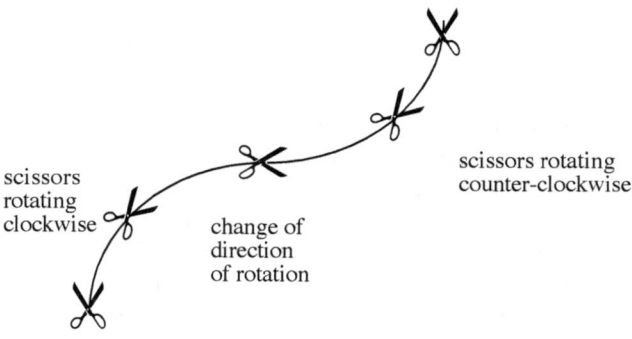

Figure 2. Point of inflection.

However, for most students, future teachers included, unless we help them develop the corresponding mathematical concepts, the symbols will remain "empty symbols, handles without anything attached, labels without contents" (Skemp 1987, p. 182). We should strive to help future teachers use their vision, spatial sense and kinesthetic sense to gain a better understanding of the ideas of calculus. In addition to unpacking each of the different conceptual approaches, future teachers need the opportunity to establish connections among them.

The number of students who take calculus in High School has increased dramatically in the last decade. In the United States the number went up from 174,000 to 325,000 between 2000 and 2010 (College Board, 2010). Given this trend of more and more students taking calculus in High School, it is very likely that most secondary mathematics teachers will teach calculus, and they need a solid conceptual understanding to teach it well. There are several classic books emphasizing a conceptual approach to calculus that can be very illuminating for secondary mathematics teachers. Some have been reprinted in recent years (Klein, 2004; Toeplitz, 2007; Courant & Robbins, 1996, especially chapters 6 and 8). Textbooks emphasizing intuitive and physical approaches are of course not new either, and in some cases are still in print (Kline, 1998; Sawyer, 1961). Textbooks that present calculus ideas numerically, graphically, symbolically, and verbally are also available (Hughes-Hallett, Gleason, McCallum, et al., 2009).

However, it is not enough that these books are available at the library cr through bookstores. Future teachers need to have the opportunity and the time to think about the concepts of calculus in the courses they take. As part of their teacher preparation program a the University of Delaware, prospective teachers take a methods course on the teaching of mathematics at the secondary level. One of the topics in that course is the teaching of calculus in High School. Future teachers spend about six hours in this course in activities, reflections, and discussions meant to give them the opportunity to engage with some of the fundamental concepts of calculus, and pause, discuss and think thoroughly about them. Most of the activities described in this chapter have been used in that course. Some of the activities were also used in a calculus course for middle school teachers at San Diego State University that relied heavily on the use of calculators and computers (Flores & McLeod, 1990, 1993). The selection of the content for both courses was influenced by theoretical considerations based mainly on research in cognition science. The courses build up abstract concepts out of more concrete examples, such as using velocity as the introduction to derivative. The goal is to improve students' ability to think conceptually about the ideas of calculus, not just to work on computational proficiency. Connections among important ideas are developed to enable students to see calculus as an integrated collection of important ideas, rather than as a disconnected set of procedural skills. The activities are conducted in class in a way that future teachers figure most of the answers on their own, working in small groups, in pairs, or individually. For the sake of completeness, most of the answers are provided. This chapter also includes a couple of examples of activities successfully used recently at other institutions by other authors.

In addition to the positive results obtained in the courses described above, other authors have documented the benefits of using an approach focusing on conceptual understanding. For example, Cummins (1977) reports that a calculus course using a discovery-experience approach to calculus, where students' initiatives, discoveries, and suggestions were encouraged and utilized to help develop the calculus, "was especially effective in promoting a deeper understanding of the calculus and that this gain was not at the sacrifice of proficiency in manipulations and applications." (p. 37)

Activities for limits of sequences

According to Cory and Garofalo (2011)

"a good understanding of limits of sequences involves a concept image of limit with the following properties: a rich example space of assorted sequences, the presence of productive conceptions of limit that are activated in the appropriate contexts, the ability to draw from the example space effectively in order to demonstrate when certain conceptions are unproductive, as well as a facility to explain correctly using words and diagrams the various parts of the definition of the limit of a sequence." (p. 68).

The examples included in this section are geared to develop such rich concept images. A sound understanding of limits of sequences can provide a foundation for developing understanding of the other concepts of calculus. A treatment of calculus based on limits of sequences rather than the epsilon - delta definition of limit of a function is given by Nitecki (2009).

Using calculators to explore limits of sequences

The following sequences can be explored by repeatedly pushing a single key after a number has been entered, or by using the number repeatedly as constant factor. Most calculators have the capability to use the same number as a constant factor. Some have a special key for that, others do not need one. The following examples will be illustrated for a calculator that remembers the last set of instructions (TI-84). The examples are used to convey the iterative process to calculate the sequences. The keys to be pushed may vary for other types of calculators.

a) Push the following keys in your calculator.
Key strokes for TI-84
2 ENTER × 2 ENTER ENTER ENTER ...
The calculator will display the numbers 2, 4, 8, 32, ...

Teachers realize that the numbers grow very quickly, the sixteenth number is already 65536. These numbers appear in the famous problem of the chessboard, or in the number of sheets when you fold a paper in half repeatedly.

b) Start with a different number. Write down your results.
What happens if you start with 1? Why?
What happens if you start with a number between 0 and 1? Try 0.1
. 1 ENTER × . 1 ENTER ENTER ENTER ...
The calculator will display the numbers
.1
.01

.001

.0001

Teachers write down the numbers on the display after each execution, and think to what number do these numbers approach. They notice that they can come as close to 0 as they want. 0 is the limit of this sequence.

Use other numbers to start with. Try 0.9

Key strokes for TI-84

. 9 ENTER × . 9 ENTER ENTER ENTER ...

The calculator will display

.9

.81

.729

.6561

Teachers write down the number on the display after each execution. They realize that in this case, too, the limit of the sequence is zero.

What keys do you have to push to obtain the following sequences?

$0.99, 0.99^2, 0.99^3, 0.99^4, \ldots$

$1.00001, 1.00001^2, 1.00001^3, 1.00001^4, \ldots$

What happens if you start with 0?

What happens if you start with a negative number? Try first a number between -1 and 0.

Try then a number less than -1.

Represent the sequence by algebraic expressions, assuming the first term is a. What can you say about the sequence

$a, a^2, a^3, a^4, \ldots$ $\quad (0 \leq a < 1)$

What can you say about the sequence

$a, a^2, a^3, a^4, \ldots$ $\quad (1 < a)$

Teachers can use a variety of initial values to obtain examples of convergent sequences that are monotonic, alternating, or constant, and of different types of divergent sequences. Further examples on the use of calculators to explore sequences and their limits can be found in Flores 1991, 1992, Flores Peñafiel 1991.

Sum of a geometric series (adapted from Brenes Castro, Díaz Campos, & Gómez Solano 1981)

Imagine you have a square cake and you divide it into three equal parts. You give one part to a friend and you keep one part. You have now 1/3 of the cake.

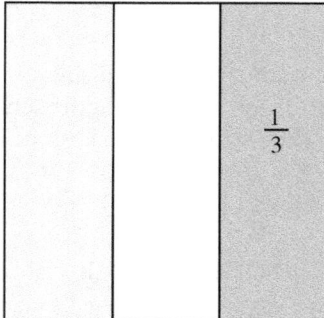

Figure 3. A cake divided in three equal parts.

Then you divide the remaining part into three equal parts. You give one part to your friend and keep one part. You have now $1/3 + 1/3^2$ of the cake.

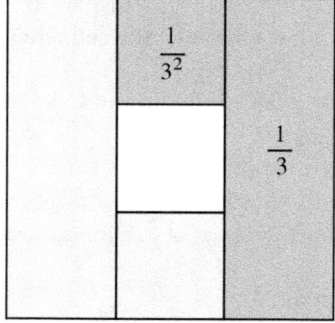

Figure 4. The remainder divided in three equal parts.

Again you divide the remaining part into three equal parts, give one part to your friend and keep one part. You have now $1/3 + 1/3^2 + 1/3^3$ of the cake.

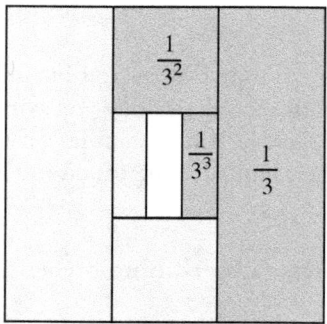

Figure 5. The process is repeated.

Repeat the process once more. You now have $1/3 + 1/3^2 + 1/3^3 + 1/3^4$ of the cake.

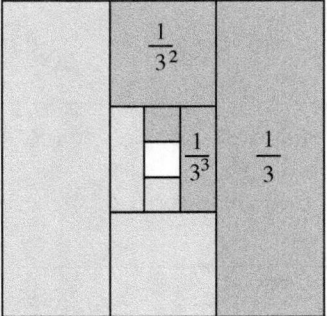

Figure 6. The process is repeated once more.

Imagine that you repeat this process an infinite number of times. How much cake will you get at the end? How much will your friend receive? What is the sum of the infinite series $1/3 + 1/3^2 + 1/3^3 + 1/3^4 + \dots$?

Small group activity. Use a similar argument to find the sum of the geometric series $1/4 + 1/4^2 + 1/4^3 + 1/4^4 + \dots$. Hint: Divide the cake into four equal parts, and share three of those parts.

Find the sum of the geometric series $1/5 + 1/5^2 + 1/5^3 + 1/5^4 + \dots$.

Find a general formula for the sum of the geometric series $1/n + 1/n^2 + 1/n^3 + 1/n^4 + \dots$

A limit with a paper strip: Approximating equilateral triangles.

This activity is adapted from Flores Peñafiel 1990 and Olson 1977.

Cut a strip of paper 5 cm wide and about 50 cm long (adding-machine paper works very well).

Fold a crease close to one end at an arbitrary acute angle A.

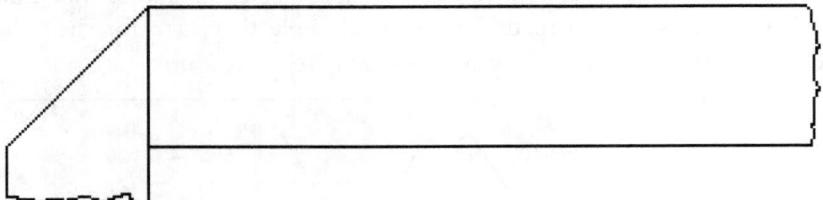

Figure 7. Folding a strip of paper.

If you unfold the strip again, you will see the acute angle A and an obtuse angle formed by the crease and the opposite edge of the strip.

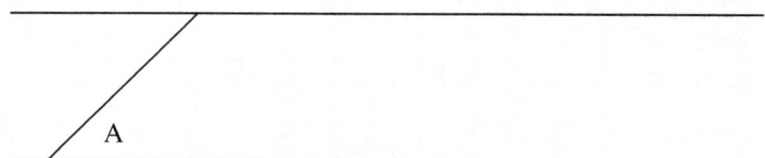

Figure 8. The first crease.

By folding, bisect the obtuse angle formed by the first crease and the edge of the strip.

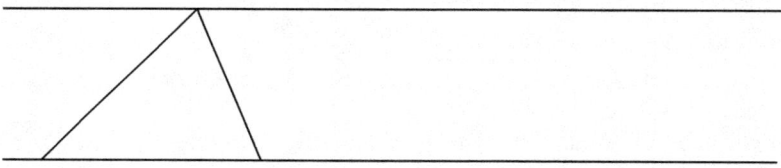

Figure 9. The second crease.

A new obtuse angle will be formed at the other edge of the strip. Bisect the new obtuse angle formed by this new crease and the edge of the strip. Continue this procedure several times. What shapes do the successive creases form?

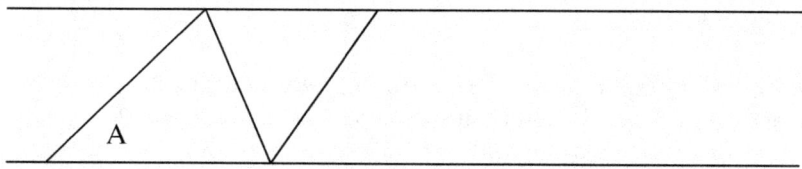

Figure 10. The third crease.

Teachers observe that after a few folds, the triangles look very much like equilateral triangles. No matter what initial angle they take for their first crease, the successive angles always approach 60° in measure.

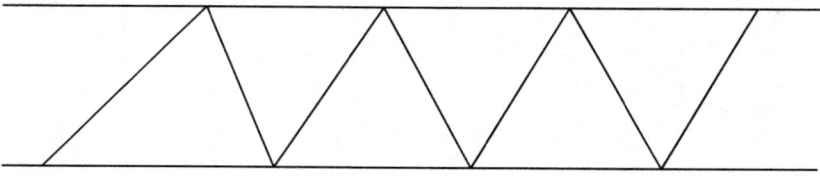

Figure 11. Zigzag pattern after seven folds.

Teachers can use their calculators to see what the successive angles are. For example, starting with an angle of 28° the sequence of key strokes on a TI-

84 corresponding to the first angle bisection would be:
28 ENTER (180 - ANS)/2 ENTER. After that, all teachers need to do is press ENTER ENTER.... to get the successive terms of the sequence. The angles would be:
28, 76, 52, 64, 58, 61, 59.5, 60.25, . . .
Teachers notice that the numbers are alternately bigger and smaller than 60, and that the difference to 60 is halved with each successive term. Then they are asked to figure out in their small groups, why this is so.
Why does it work? Folding the strip of paper as indicated will form an obtuse angle that is (180 - A), bisecting it will give an angle $A_2 = (180 - A)/2$, as indicated in Figure 12.

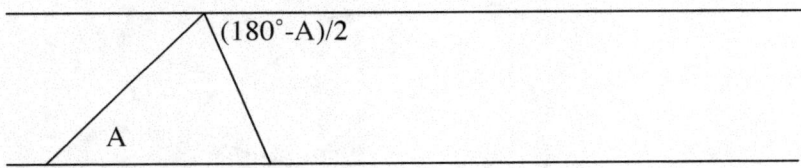

Figure 12. Bisecting the obtuse angle.

The following angle will be $A_3 = (180 - A_2)/2$. So that in general
$A_n = (180 - A_{n-1})/2$ $A_1 = A$ $(0 < A < 180)$
Let E be the difference between the arbitrary initial angle A and 60. If we express the angle A as $60 + E$, we can see that the next angle will be
$(180 - (60 + E))/2 = (180 - 60 - E)/2 = (120 - E)/2 = 60 - E/2$. Thus the difference has been halved, and it will be of opposite sign. That is, if the first angle is bigger than 60, the next will be smaller than 60, but closer to 60 (the distance is half as big as before).

Two limits related to the circle
Area of a $2n$-sided polygon and its relation to the perimeter of a n-sided polygon.
As a preliminary activity, teachers are asked to remember or figure out the formula for the area of a kite in terms of its diagonals. Area kite $= \frac{1}{2} d_1 d_2$.

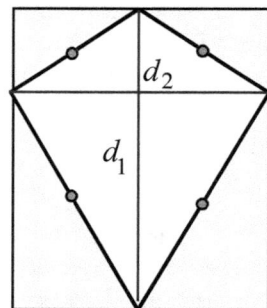

Figure 13. Area of a kite.

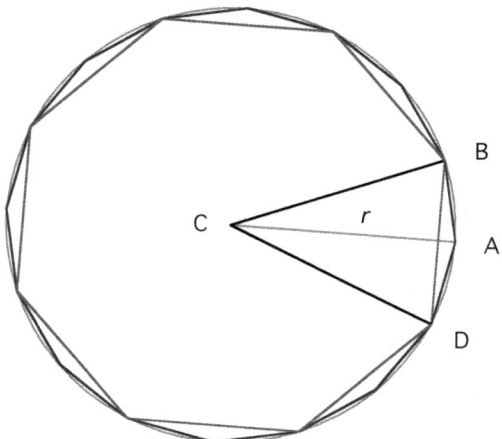

Figure 14. Regular polygons inscribed in a circle.

Teachers are given a diagram with two regular polygons inscribed in a circle (Figure 14, adapted from Nelsen 2000, p. 23). One polygon has twice as many sides as the other. Teachers are asked to express the area of kite ABCD in terms of its diagonals, one of which is the radius and the other one side of the n-sided polygon.

By breaking the $2n$-sided polygons into n kites, teachers find out that the area of the $2n$-sided polygon is equal to one half of the product of the radius by the perimeter of the n-sided polygon, $A_{2n} = rP_n/2$. By increasing the number of sides, the perimeter of the n-sided polygon will approximate better and better the circumference of the circle and the area of the $2n$-sided polygon will approximate the area of the circle. Thus

$$\lim_{n\to\infty} A_{2n} = \lim_{n\to\infty} \frac{1}{2}rP_n = \frac{1}{2}r \lim_{n\to\infty} P_n = \frac{1}{2}r \times 2\pi r = r^2\pi.$$

This activity helps future teachers understand better why the same constant

π appears in both formulas for the area of a circle and the perimeter. An additional insight that can be gained by computing the area of the $2n$-sided polygon in terms of the perimeter of the n-sided polygon is that there is a practical difference between approximating π by using areas of inscribed polygons or the perimeters. To get the same degree of precision we will need twice as many sides if we use area than if we use perimeter (Flores, 2002).

Understanding limits of sequences with an epsilon strip

Recent research (Cory and Garofalo 2011, Roh 2008) shows that teachers and students have problems understanding the formal definition of limit of a sequence. This adds to the body of research that documents that students have many difficulties with the concept of limit, and that it is hard to redress some of their misconceptions (Cornu, 1991; Cottrill, Dubinsky, et al., 1996; Davis & Vinner, 1986; Hitt & Páez, 2001; Vinner, 1991; Williams, 1991). The formal definition of limit of a sequence is $L = \lim_{n \to \infty} a_n$ if and only

if for all $\varepsilon > 0$ there is an $N \in \mathbb{N}$ such that for all $n \geq N$, $|a_n - L| < \varepsilon$. This definition can be quite intimidating and it involves several different variables. Students are often not aware what is the exact role that the different variables play in this definition. Fortunately, several authors (Barnes, 2011; Cory & Garofalo, 2011; Roh, 2010) have also designed and tried two-dimensional, dynamical ways to help develop a better understanding of the formal definition of the limit of a sequence. As students use a mobile strip of width epsilon, they can see better the role of N, and how in general it depends on how small epsilon is. Cory (2009) used a dynamical geometry program to develop interactive epsilon strip sketches. Students can determine the possible value of the limit by dragging, and they can change the value of epsilon and N and see what points of the sequence fall within the strip. Cory includes graphs of a monotonically decreasing sequence, a damped oscillating sequence, a non-convergent oscillating sequence, and a constant sequence, among others. One method Cory uses to help student understand the formal epsilon-N definition of limit of a sequence is using the epsilon-N game. The instructor sets the red epsilon slider to a particular value, and a student is asked to slide the green N line along the x-axis and try to find a value of N beyond which all further terms are within the epsilon band (see Figure 15).

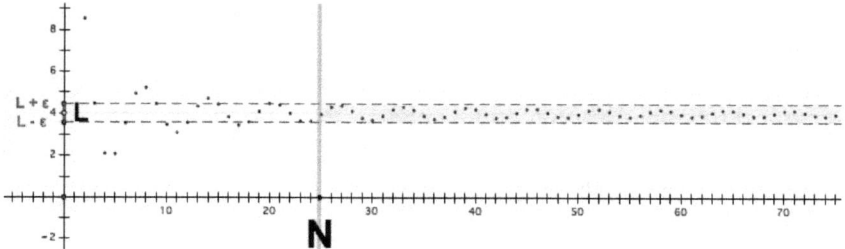

Figure 15. An epsilon strip.

Cory studied three pre-service secondary teachers development of their understanding of limit of sequence by using this approach (Cory and Garofalo, 2011). Participants manipulated the external visual representation of the formal limit concept in the contexts of various sequences. As a consequence of the interactions with the software they seemed to strengthen their connections between their visual models and the verbal representation. Cory and Smith (in press) used this approach in conjunction with rich mathematical discussion successfully to enable college calculus students construct the formal definition of limit of a sequence on their own.

Activities for the derivative and related concepts

Method of Increments

Teachers were asked to unpack the meaning of the quotient of increments $\frac{f(x+h)-f(x)}{h}$ both in the context of secant lines to the graph of the function and in terms of average speed of an object moving with variable velocity. The slope of a secant line to the graph of a function f, with a specific increment h, is given by the quotient of increments $\frac{f(x+h)-f(x)}{h}$. If the function represents distance and the independent variable represents time, then the slope of the secant represents the average speed of the object in the given time interval.

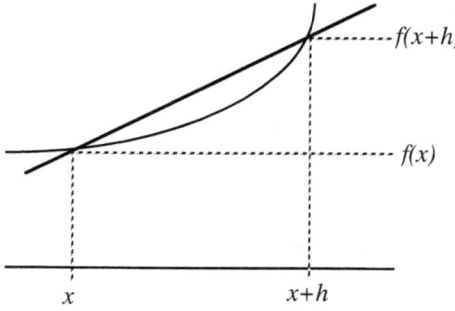

Figure 16. The slope of a secant line.

Then teachers were asked to discuss the situation of having successive quotients of increments determined by smaller and smaller increments h, both in the context of secant lines and average speed. The geometrical idea behind the method of increments is that to find the tangent line to a curve on a given point you can use a sequence of secant lines to that curve, passing through the given point and a second point on the curve. The slope of the secant will approximate the slope of the tangent line if the second point of intersection of the secant line is close enough to the given point (Figure 17). In terms of velocity, the value of the average speed will approximate the instantaneous speed at the given time, as the time increments get smaller and smaller.

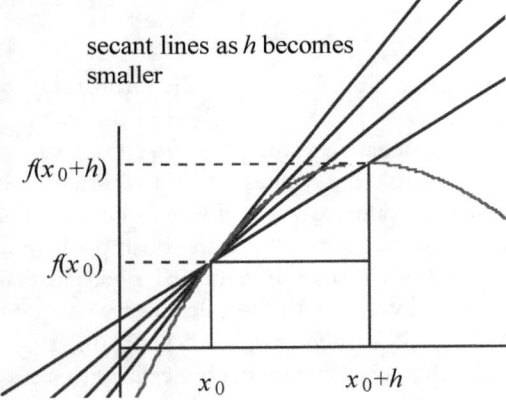

Figure 17. Secants with smaller increments.

The method of increments is of course not new (see Kline, 1967 p. 21; Courant and Robbins, 1941; Sawyer, 1961). However, often future teachers do not have the opportunity to pause and reflect on each of the parts of this method, first the quotient of increments and then the limit. For example, after doing the activities related to quotient of increments, one future teacher wrote in her reflection, "I know that I personally can't remember a time when I was a high school, or even college, student, that the quotient of increments was ever really properly explained to me." (Allie Rodriguez, Fall 2010)

A calculator program to compute the slopes of secants
For this activity related to the method of increments, teachers use a short program for the TI-84 calculator to compute the slope s of the secant to the graph of any function $y(x)$, passing through a given initial point $(x, y(x))$ with a given initial increment h. The calculator will then automatically take smaller and smaller values of h. In the $Y=$ menu teachers can enter any

function they want in the $Y_1=$ slot and the program will use that function. So every time teachers want to run the program with a new function all they need to do is change the function defined in Y_1.

```
PROGRAM SLOPSECF           name of program
Prompt X                   enter given value of x
Prompt H                   enter increment
For(N,1,6)                 start of cycle, calculator will do six times
(Y1(X+H)-Y1(X))/H→S        computation of slope of secant line
Disp H,S,""                display increment and slope
Pause                      pause, press ENTER to continue
H/10→H                     smaller value of increment
End                        end of cycle
```

Teachers try the program first for $y = x^2$. They run the program for $x = 2$ and an initial value of $h = 1$. Teachers observe the values of s. Then they run the program for a negative value of h, first for an initial value $h = -1$. They observe and discuss the values of s. In either case, h positive or negative, they realize the values of s get closer and closer to 4.

Then they repeat the activity for $x = 1$ using both positive h and negative h. They observe that in both cases the values of s get closer and closer to 2. Teachers conjecture and verify with the calculator what the relation of s and x is for different values of x, such as 3, 4, 0, -1, -2, 10, -10.

Finally they describe this relationship algebraically with an equation, $s = 2x$. Teachers then modify the function in the $Y_1=$ slot to obtain the slopes of the secants to the function $f(x) = 2x^2$ that go through a given arbitrary point x, and repeat the whole activity. They describe the relationship between x and s with an equation, $s = 4x$. Teachers can then conjecture and generalize what the relationship between x and s will be for general functions of the form $y = ax^2$.

The derivative of the area of the circle

In this activity teachers are asked to use the method of increments on the area of the circle to study the relationship between the area of a circle and its circumference from a different perspective. The increment of the area of a circle when its radius r is incremented by h is the area of the ring of width h, $A(r+h) - A(r) = \pi(r+h)^2 - \pi r^2 = \pi(2r+h)h$ (Figure 18). Because $2r+h$ is the average of the diameter of the original circle and the diameter of the incremented circle, teachers can see that this area is given by the product of the length of the red circle and h (Figure 19). The ratio of increments $\frac{A(r+h)-A(r)}{h} = \frac{\pi(2r+h)h}{h} = \pi(2r + h)$ is thus equal to the circumference of the red circle (Figure 19). As h tends to zero, this circumference will tend to

$2\pi r$. Therefore, the derivative of the area of a circle with respect to the radius is the circumference of the circle. Although of course teachers were able to obtain the derivative of the area with respect to the radius, $\dfrac{d(\pi r^2)}{dr} = 2\pi r$ before doing this activity, they can appreciate the additional insight obtained by using the method of increments.

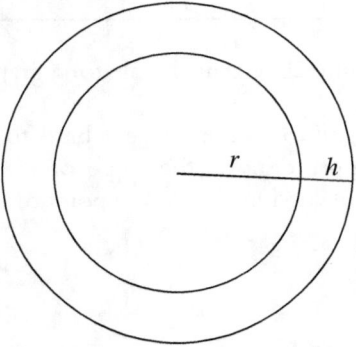

Figure 18. Increment of area.

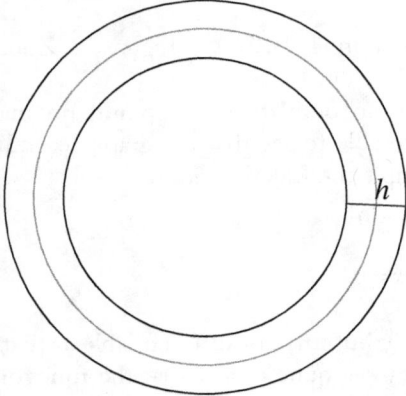

Figure 19. Ratio of increments equal to mid circumference.

The derivative and locally straight graphs

Zooming-in on a graph using graphing calculator or a computer software like *GeoGebra* or *Graphing Calculator* (Pacific Tech, 2007) can help students see that functions that have a derivative at a point look locally like a straight line under enough magnification. Future teachers were asked to zoom in on points where the derivative did exist, for example (1, 1) for the function $y = x^2$.

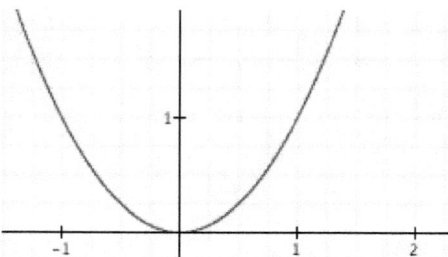

Figure 20. Original scale for a graph.

After a couple of magnifications, it is very hard to distinguish the graph from a straight line. Furthermore, the slope of the "straight line" is the same as the derivative of the function at the point of interest.

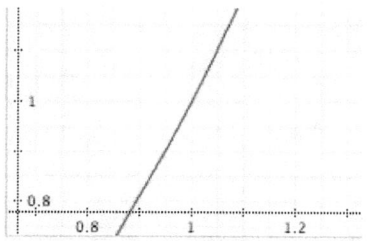

Figure 21. Zooming in on (1, 1).   Figure 22. Zooming in some more.

However, if there is no derivative at a point, no matter how much we magnify, we won't be able to see that the graph resembles a straight line at such point. For example the function defined as

$$\begin{cases} y = 0 & \text{if } x = 0 \\ y = x\sin\left(\dfrac{1}{x}\right) & \text{if } x \neq 0 \end{cases}$$

is continuous at $x = 0$, but it is not differentiable at that point. When future teachers zoom-in they see quite clearly that the function is not at all locally straight at 0.

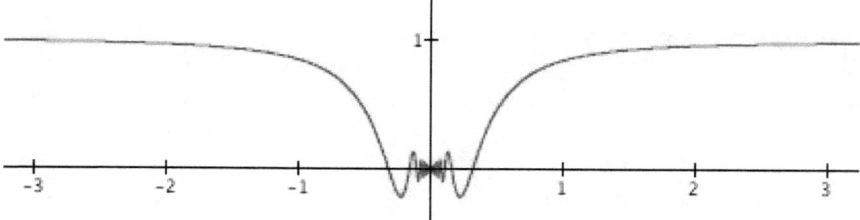

Figure 23. Graph of $y = x \sin (1/x)$.

30

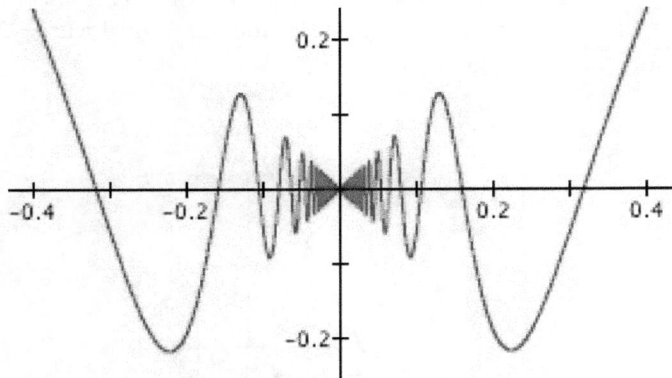

Figure 24. Zooming in at zero.

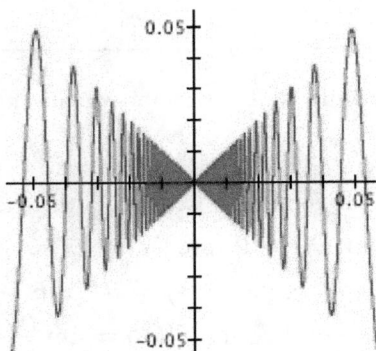

Figure 25. Graph is not locally straight at zero.

Derivative and best linear approximation

A complete understanding of the derivative is not possible if we ignore the connections of the derivative to other concepts in calculus. In this section we will discuss the relation of the derivative with the linear approximation and the differential.

The derivative allows us to find the tangent to a curve at a given point. With this we can approximate locally a differentiable function by means of a linear function. Let us consider the situation where x is a fixed value and h is variable. Because $f'(x)$ is the tangent of angle α in Figure 26, $f'(x)h$ represents the opposite leg of the right triangle. We see that the difference $f(x+h) - f(x)$ is equal to the sum $f'(x)h + \varepsilon(h)$. The main part of the increment of the function will be given by the increment $f'(x)h$, which is proportional to h, that is, $f'(x)h$ is a linear function of h. This linear function is the differential. For example, if $f(x) = x^2$, the differential will be $2xh$, and the error in this case is given by $\varepsilon(h) = h^2$.

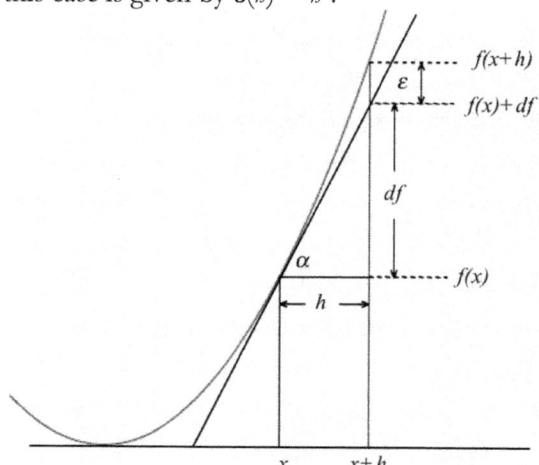

Figure 26. Linear approximation.

The part $\varepsilon(h)$ not only tends to zero as h tends to zero, but we can make $\varepsilon(h)$ as small as we wish compared with h, if we take h small enough. In more precise terms, $\lim_{h \to 0} \frac{\varepsilon(h)}{h} = 0$. Students can see this in the following way. From the equation $f(x+h) - f(x) = f'(x)h + \varepsilon(h)$ we have that $\frac{f(x+h)-f(x)}{h} = f'(x) + \frac{\varepsilon(h)}{h}$. Because $f'(x) = \lim_{h \to 0} \frac{f(x+h)-f(x)}{h}$ $f'(x) =$, we have $\lim_{h \to 0} \frac{\varepsilon(h)}{h} = 0$. Thus we see that we can approximate $f(x+h)$ by means of $f(x) + f'(x)h$, and that for small values of h the error is very small compared with h.

Error with linear approximation
If we know the value of a function $y = f(x)$ and its derivative $f'(x)$ at a point x, we can compute linear approximations to the values of the function at points $x + h$ that are close. If the value of h is small, then the error e will be small even compared with h. Notice that $df = f'(x)*h$ is the linear increment. The following program for TI-84 will compute the linear error, and compare it to the increment. In slot Y1 of the Y = list enter the desired function. In slot Y2 enter the derivative of the function. For example Y1 = X * X, Y2 = 2X

PROGRAM:ERRLIN

```
Prompt X
Prompt H
For(N,1,6)
Y₁(X+H)-Y₁(X)-Y₂(X)*H → E
Disp H,E,E/H,""
Pause
H/10→H
End
```

Run the program to compute the error of linear approximation to $y = x^2$, for $X = 2$ and $H = 1$. The program will pause after displaying a set of values H, E, E/H; press ENTER to continue. Notice that for small values of H the error E is much smaller than H, and that $E/H \to 0$ as $H \to 0$.

Now run the program for $H = -1$. How is the size of the error compared to the increment H?

Run the program for other values of X and for increments H that are positive and negative.

Change the functions in Y1 and Y2 to obtain linear approximations to $y = x^3$.

Y1 = X * X * X
Y2 = 3 * X * X

Run the program for different values of x and for increments h that are positive and negative.

Run the program for other functions. Write the new function in Y1 and its derivative function in Y2.

The Babylonian method for square root. Why is it so efficient?

Another activity designed to help teachers make explict the ideas of derivative and linear approximation is have them look at the ancient Babylonian method to approximate square roots.

The Babylonian method to approximate the square root of a number through successive approximations is remarkably efficient. Using modern notation we can describe the method in the following way. To obtain the square root of a number n, give a first approximation a_1. Then divide n/a_1, which will also give you an approximation. Finally compute the average of the two values: $(a_1 + n/a_1)/2$. This will give a second approximation a_2 which is much better than the first. Repeat the process to obtain the desired precision. Thus, we obtain recursively a sequence of approximations

$$a_{m+1} = \frac{1}{2}\left(a_m + \frac{n}{a_m}\right).$$

With a calculator obtain the square root of 5 using this procedure. Use 2 as your initial guess. How many times do you need to repeat the process to get all the precision allowed by the calculator?

With a TI-84 calculator you can press the following keys
2 ENTER (2ND ANSWER + 5 ÷ 2ND ANSWER) ÷ 2 ENTER
ENTER ENTER

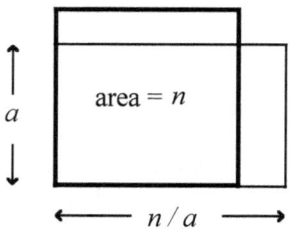

Figure 27. Error in opposite directions.

Arithmetic analysis. Compare the errors involved in the first step.

$\sqrt{5}$ - 2 = .23067977

$\sqrt{5}$ - 2.5 = -.263932022

Notice that the difference from the first approximation 2 to $\sqrt{5}$ is very close to the difference from 5/2 to $\sqrt{5}$ but with opposite sign (see also Figure 27). Therefore, when averaging the two values we can expect a much better approximation. We can see that the errors involved in the second step are much smaller.

$\sqrt{5}$ - 2.25 = -.013932022

$\sqrt{5}$ - 5/2.25 = .0138455755

In the example above, our first approximation a was below the value of the square root and when dividing n/a we obtained a number that is bigger than the square root. Let's look at an example where the first approximation is above the square root. In that case when dividing n/a we will get a number that is smaller than the square root.

Approximation to $\sqrt{80}$ using the Babylonian method

$$\frac{9+\frac{80}{9}}{2} = \frac{81+80}{2\times9} = \frac{161}{18}.$$

To approximate the square root of 80 we can use also the differential. $\frac{d\sqrt{x}}{dx} = \frac{1}{2\sqrt{x}}$ and take $dx = -1$ So the linear approximation to $\sqrt{80}$, starting with $\sqrt{81} = 9$ is $\sqrt{81} - 1 / 2\sqrt{81} = 9 - 1 / (2\times9) = 161/18$.

We see that using the differential gives us the same result as using the Babylonian method.

Analysis of the linear approximation

$r = a_1 + e$

$r^2 = (a_1 + e)^2$

$n = a_1{}^2 + 2\,a_1 e + e^2$

$n \approx a_1{}^2 + 2a_1 e$

$e \approx (n - a_1{}^2)/2a_1$

Notice that in Figure 28, $\tan \alpha = 2a_1$

Let $e_1 = (n - a_1{}^2)/2a_1$

$a_2 = a_1 + e_1$

$a_2 = a_1 + (n - a_1{}^2)/2a_1 = (a_1 + n/a_1)/2$

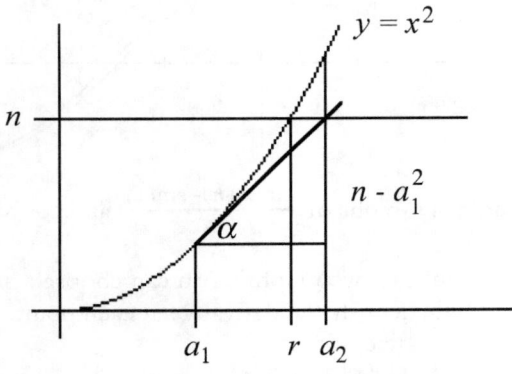

Figure 28. Linear approximation to the square root.

From derivative at a point to the derivative function.

As mentioned above, a subtle difference in notation often masks the fundamental difference between derivative at a point and the derivative function. In one case, we deal with an arbitrary but fixed point, in the other with a variable. In one case the result is a number, in the other the result is a function. In order to give teachers the opportunity to unpack what is meant to use quotients of increments when x is a variable, they approximate the derivative functions with families of functions defined as quotients of increments.

Approximating the derivative function

For a given point x_0, $\lim\limits_{h \to 0} \dfrac{\sin(x_0 + h) - \sin(x_0)}{h}$ is the derivative at the point x_0.

For a differentiable function like $\sin(x)$, the quotient of increments $\dfrac{\sin(x_0 + h) - \sin(x_0)}{h}$ gives a pretty good approximation to the derivative at the point x_0 for h sufficiently small. Now we can imagine that we approximate simultaneously the derivative at all points. We will use a graph generated by the *Graphing Calculator* program.

The purple graph is $y = \frac{\sin(x+h)-\sin(x)}{h}$ for a value of $h = .2$ and the blue graph is $y = \cos(x)$. *GeoGebra* and *Graphing Calculator* allow students to use h in a slider, so that they can see how for smaller values of h, the approximation is getting better.

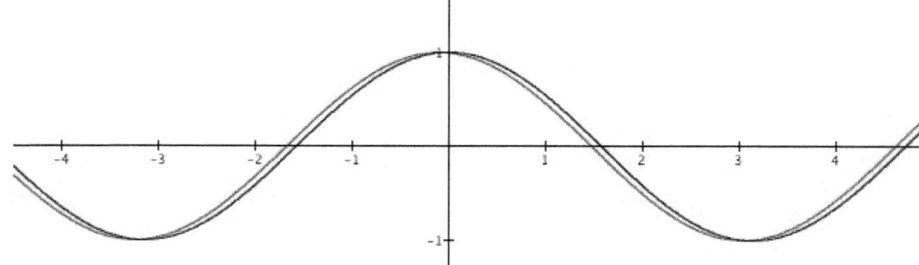

Figure 29. Graphs of $\frac{\sin(x+h)-\sin(x)}{h}$ and $\cos(x)$.

What can you say about the two graphs? You can choose a smaller value of h to approximate simultaneously the derivative at each point.
Approximating the derivative of $\tan x$.
The purple graph is $y = \frac{\tan(x+h)-\tan(x)}{h}$ for a value $h = .1$; the blue graph is $y = sec^2(x)$.

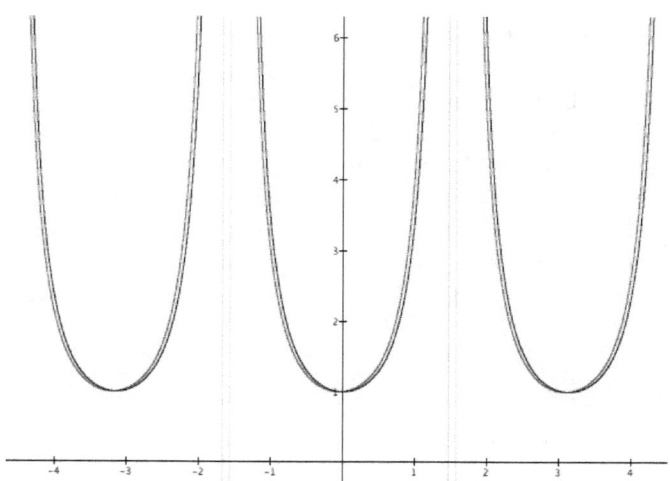

Figure 30. Graphs of $\frac{\tan(x+h)-\tan(x)}{h}$ and $sec^2(x)$.

Interlude: Motion activity

One useful activity in the methods course, that is not part of the calculus sessions, is to have future teachers use a motion detector to match first a time vs. distance graph and then a time vs. velocity graph generated by the graphing calculator (piecewise linear functions). Teachers are the moving objects as they walk towards and away from the motion detector. With the time vs. distance graph they usually do not have any problems interpreting the different parts of the graph, and adjusting their motion to match the graph. It is interesting, however, that with the time vs. velocity graph, some misinterpretations and difficulties occur. Future teachers have to think through what the different parts of the graph represent. It also becomes clear for them, when they try to match the velocity graph, that the initial distance to the detector does not matter, just how fast they move and the direction of the motion.

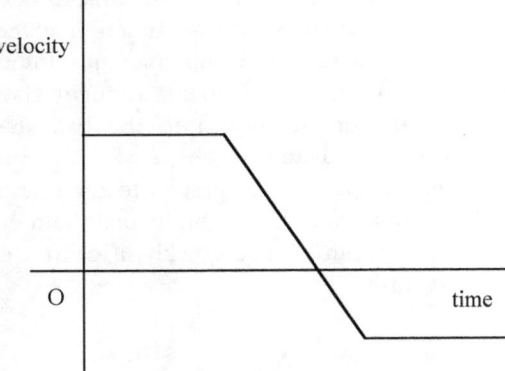

Figure 31. Time vs. velocity graph.

Derivative functions using a motion detector

Cory (2010) conducted an intense problem-solving activity with her calculus students using bouncing balls and motion detectors. Students first observed a bouncing ball dropped by the instructor and graphed the height of the ball vs. time. An interesting discussion ensued whether the graph had sharp cusps on the x axis, as most of the graphs made by students showed. Then students obtained the graph of the bouncing ball using a motion detector.

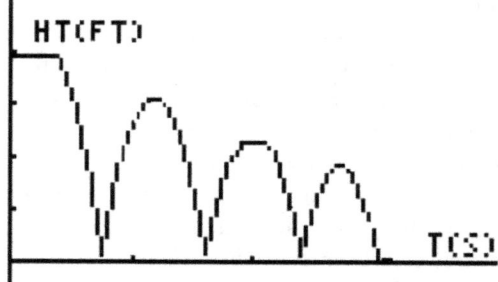

Figure 32. Height of a bouncing ball (Cory, 2010).

Students were asked then to predict what the graph of the velocity vs. time would look like, that is, the graph of the derivative function of the height. They used their previous knowledge of the relationship between the values of the derivative and the behavior of the function at certain points or intervals, such as local maxima and minima, and intervals where the function is decreasing or increasing. Students thought about the shape of the graph in relation to what happened to the ball at different times, discussed with each other, predicted the shape of the graph of the velocity and drew it. Then they compared their graphs to the one produced by the motion detector. The activity and discussion helped them understand better how, although the velocity can change quickly, it cannot change direction abruptly and instantaneously.

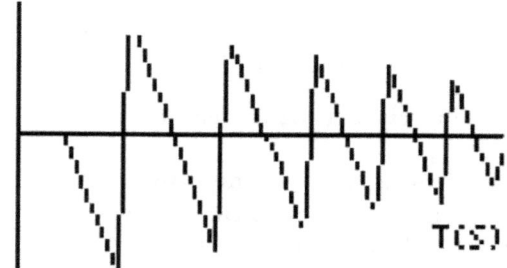

Figure 33. Velocity of bouncing ball (Cory, 2010).

Introduction to the integral: Motion and area

Distance in a time-velocity graph.

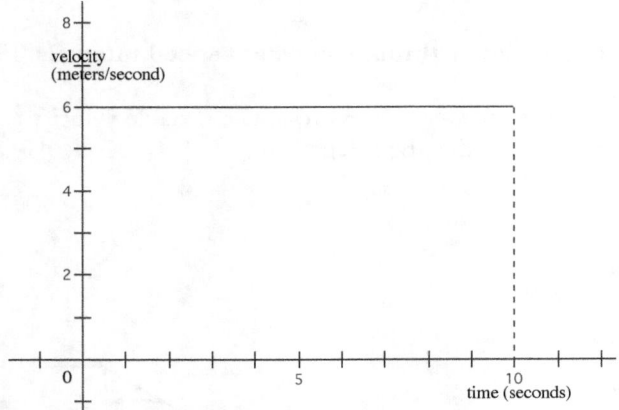

Figure 34. Graph of uniform speed and distance as area

The area under the graph is equal to the product of its length by its width. The length of the rectangle is given by the time, and the width is given by the velocity. So the area of the rectangle is given by *vt*. But from the formula for moving bodies $d = vt$, this product is also the distance traveled, so that an interpretation for the area under a graph of constant speed is the distance traveled.

Small group task. Design a lesson to help students understand that the area under the graph of a uniformly accelerated motion is also the distance traveled by the object. Several approaches are possible. Galileo compared the distance traveled by an object with uniform acceleration to the distance traveled with an object at the average speed.

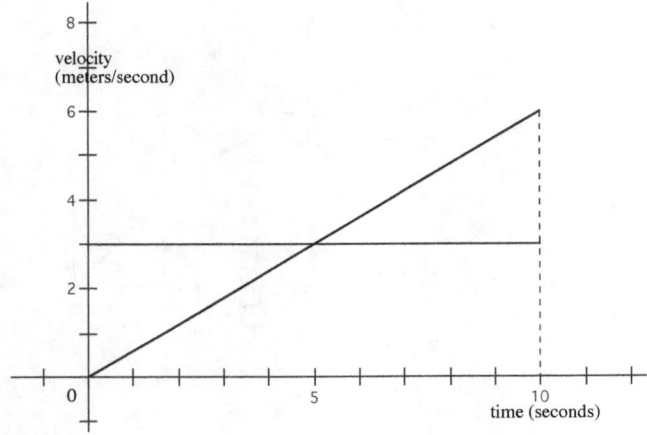

Figure 35. Distance at average speed.

39

Approximating distance through constant speed intervals (Fremont 1978, Ch. 9)

The graph in Figure 36, $v = 3/5\ t$ represents a variable velocity graph. What do the rectangles under the graph represent?

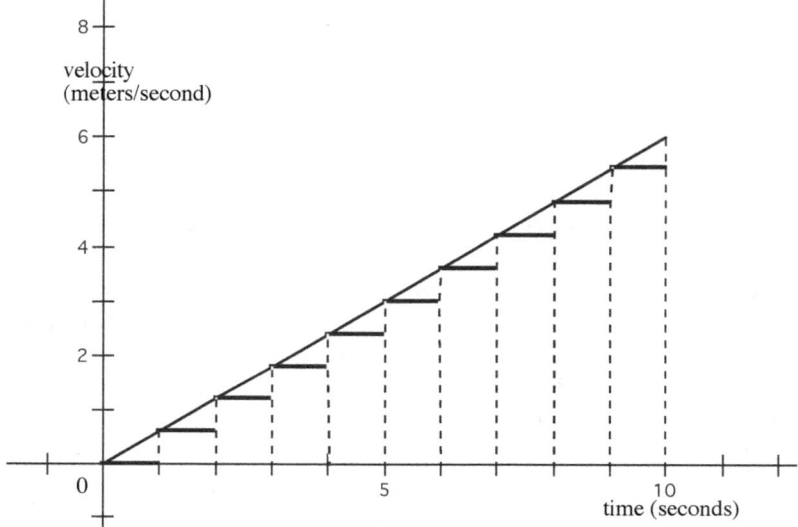

Figure 36. Approximation by a step function

Discuss in your group how you could get an even better approximation to the distance traveled by the object.

Area under a curve

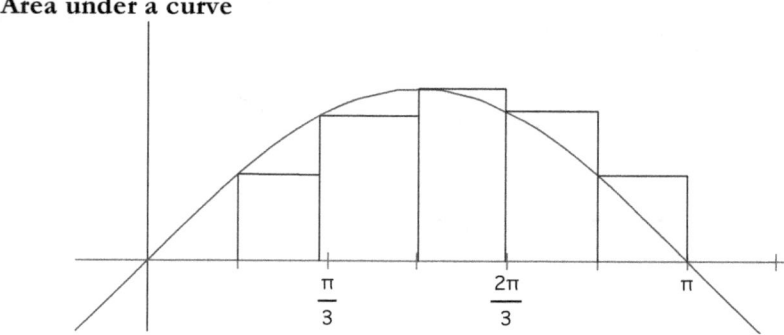

Figure 37. Approximation to the area under the curve between 0 and π.

Discuss in your group how you could approximate the area under a curve

40

for a given interval by using rectangles. Draw the picture of a graph and an approximation to the area by using rectangles. In your picture make explicit each of the elements of a Riemann sum $\sum_{k=1}^{n} f(x_k)\Delta x_k$. Discuss how you could get an even better approximation.

Teachers used two methods to obtain better approximations. One was to increase the number of rectangles by making their bases Δx_k smaller. The other was to use trapezoids instead of rectangles, which can be used to illustrate the idea behind different numerical methods used in practice.

Let f be a continuous function. Discuss in your group what the limit

$$\lim_{n\to\infty} \sum_{k=1}^{n} f(x_k)\Delta x_k \text{ represents.}$$

Logarithms

The natural correspondence between a geometric progression and an arithmetic progression permits us to add instead of multiply. According to Infeld (1978), when Galois took his examination to enter the Ecole Polytechnique, one of the questions was about logarithms. Galois simply wrote two progressions on the blackboard, an arithmetic progression and a geometric one in a correspondence like this

a) 0   1   2   3   4   ...
b) 1 a a^2 a^3 a^4 ...

The terms of the arithmetic progression are the logarithms of the geometric progression and the base is a. To multiply the terms of the geometric progression it is sufficient to add the corresponding terms in a) and find the term in b) that corresponds to the sum. This is the result of the multiplication. Thus we can multiply any two terms of the geometric progression. For example, with the progressions

a') 0   1   2   3   4   5   6   7   8   9   ...
b') 1   2   4   8   16   32   64   128 256 512 ...

to multiply 16×32 we add $4 + 5$ and look under 9 to find the result of the multiplication, 512.
But what if we want to multiply numbers that are not in the progression? The terms in b') are too far apart. We can think of choosing for the base a number closer to 1 than 2, for example 1.0001, if we are interested in numbers with up to four digits beyond the decimal point. This is the approach that Bürgi took to logarithms, which differs from Napier's (see

Voellmy 1974, Goldstine 1972).

a'')	0	.0001	.0002	.0003	.0004	.0005	...
b'')	1	1.0001	1.0001^2	1.0001^3	1.0001^4	1.0001^5...	

This provides with a sequence of number that are closer together, 1, 1.00020001, 1.00030003..., 1.00040006..., 1.00050010..., ...

However, as we advance in the sequence, the distance between the terms in the geometric sequence will increase gradually
... 5.1465..., 5.1472..., ...

Eventually, the gaps will be too big for the desired precision. We could think of another number for the base that is even closer to 1, like 1.00000001

0	.000000001	.00000002	.00000003	...
1	1.00000001	1.0000001^2	1.000000001^3	...

This would give us the desired precision for a longer time, but eventually again the gaps will become too big. To avoid the problem we need a different approach, an approach that will allow us to multiply any two numbers. The following geometric approach is based on the discussion on logarithms and the exponential function by Felix Klein (Klein, 2004).

Consider the following progressions, one arithmetic and the other geometric

c)	0	$1/n$	$2/n$	...	1	...	m/n	...
d)	1	$1+\dfrac{1}{n}$	$\left(1+\dfrac{1}{n}\right)^2$	...	$\left(1+\dfrac{1}{n}\right)^n$	...	$\left(1+\dfrac{1}{n}\right)^m$	...

If we draw the points of d), the geometric progression, on the horizontal axis, it looks something like this:

$$1 \quad (1+1/n) \qquad\qquad (1+1/n)^{\wedge}m$$

Figure 38. A geometric sequence on the number line.

The lengths of the intervals form a geometric sequence

$$(1)\quad \frac{1}{n}, \quad \frac{1}{n}\left(1+\frac{1}{n}\right), \quad \frac{1}{n}\left(1+\frac{1}{n}\right)^2, \quad \frac{1}{n}\left(1+\frac{1}{n}\right)^3, ...$$

We will draw on each interval a rectangle of area $1/n$, that is, the distance between the terms of c), the arithmetic sequence. To do this, we need to draw rectangles of heights equal to

$$1, \quad \left(1+\frac{1}{n}\right)^{-1}, \quad \left(1+\frac{1}{n}\right)^{-2}, \quad \left(1+\frac{1}{n}\right)^{-3}, \ldots.$$

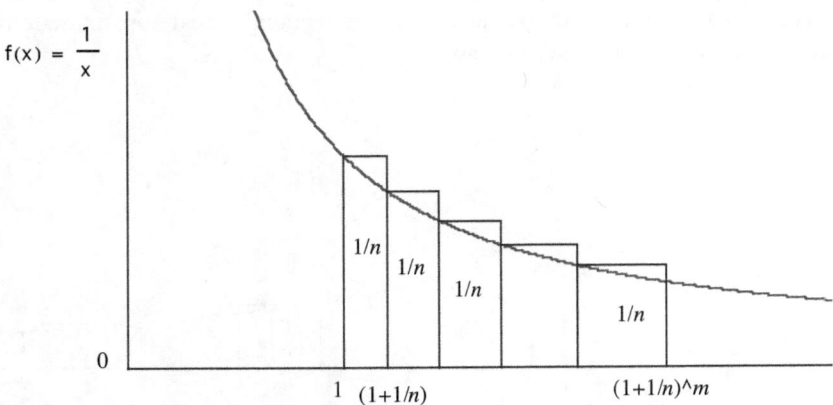

Figure 39. Equal area rectangles.

The values of the heights are the reciprocals of the values in d). That is, one of the vertices of each rectangle is on the graph $y = \frac{1}{x}$.

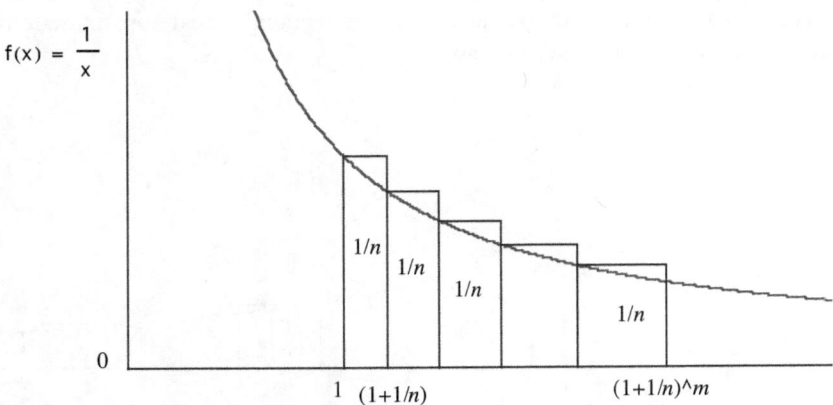

Figure 40. The function $f(x)=1/x$ and rectangles with equal areas.

The sum of the areas of the rectangles from the first to the m-th is m/n and this is precisely the term that corresponds to $\left(1+\frac{1}{n}\right)^{m}$ in the progression c).

An interesting term in the geometric sequence is $\left(1+\frac{1}{n}\right)^{n}$, the term that corresponds to an area of 1.

The geometric interpretation of the correspondence between the arithmetic and the geometric sequences suggests that in order to be able to assign logarithms to any number, and not just numbers on the geometric sequence, we can use the area under the curve $y = 1/x$. We make correspond to a number a the area under $y = 1/x$ between 1 and a. All we need to do is to verify that indeed this correspondence does have the property that we want, that for any two positive numbers a and b, the area between 1 and ab, is the same as the area between 1 and a plus the area between 1 and b. Or using the language of calculus, we need to verify that

$$\int_1^{ab} \frac{1}{x}\, dx = \int_1^{a} \frac{1}{x}\, dx + \int_1^{b} \frac{1}{x}\, dx$$

To see that this is true, all we need to do is show that $\int_1^{b} \frac{1}{x}\, dx = \int_{a}^{ab} \frac{1}{x}\, dx$ (1).

A specially interesting number is the one that corresponds to an area of 1. We call this number e.

Exercise. Give a geometric interpretation of equation (1) and prove it.
Hint: Partition the interval $(1, b)$ and construct rectangles to approximate the area under the curve, and then multiply each of the points of the partition by a. This will give you a partition of the interval (a, ab). Show that the corresponding rectangles that approximate the area under the curve between a and ab, have the same area as the rectangles that approximate the area under the curve between 1 and b.

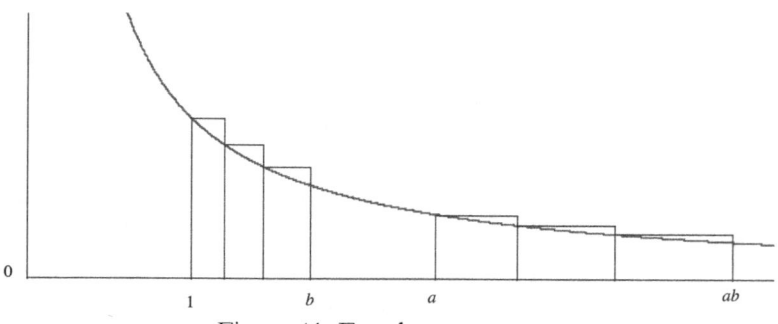

Figure 41. Equal areas.

Thus, if we define $\log (a) = \int_1^{a} \frac{1}{x}\, dx$ we have then the properties

$\log (ab) = \log (a) + \log (b)$
$\log (1) = 0$
$\log(1/a) = -\log(a)$

The function log is thus an isomorphism between two groups, the group of positive real numbers with multiplication, and the group of all real numbers with addition.

The indefinite integral and the fundamental theorem of calculus

From definite to indefinite integral

As was the case with the concepts of derivative at an point and derivative function, where the difference of notation is very subtle, but the concepts are very different in nature, the distinction between definite integral and indefinite integral is also very subtle in notation, but conceptually they represent very different kinds of mathematical objects. The definite integral is a number, while the indefinite integral is a function.

Teachers are asked to remember the definition of definite integral of a function $\int_a^b f(x)\,dx = \lim_{n\to\infty} \sum_{k=1}^n f(x_k)\Delta x_k$, and to discuss what does the definite integral mean and what it represents geometrically.

Then they are asked to discuss what the indefinite integral means $\int_a^x f(t)\,dt$, and what kind of mathematical object is the indefinite integral.

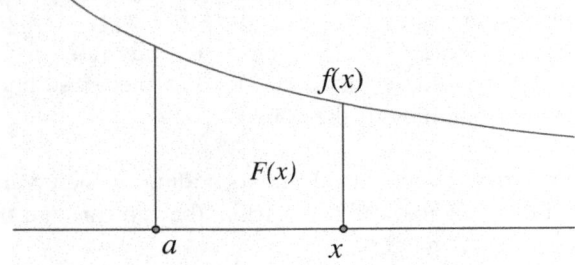

Figure 42. Area under a curve as a function of x.

One way to develop an understanding of the indefinite integral is to think about the area under the graph of a function from a fixed point a to a variable point x, so that the area changes as x changes along the horizontal axis (Figure 42). Husch (2001) provides an applet where students can drag x, and the computer will show simultaneously the swiped area and the corresponding graph for the values of $F(x)$ (Figure 43).

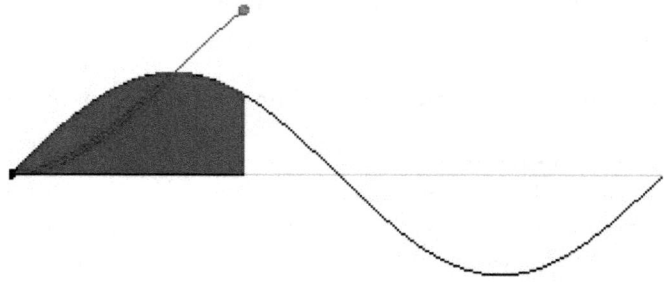

Figure 43. Changing area under the curve.

Fundamental theorem of calculus
One of the crucial insights in calculus is the inverse nature of the derivative and the integral. However, students often are presented with the fact that in order to obtain an integral it is enough to find an antiderivative, with no explanation of why this is true. This approach masks the fact that the integral can be defined independently of the concept of derivative and the profound nature of the fundamental theorem of calculus. One student in the methods course made a comment about the inverse nature of derivative and integral in his written reflection: "After taking calculus, numerous people I've spoken to still don't fully understand why there is an inverse relationship. They can do the computations, but they have no idea as to why the computations actually work out the way they do" (Trevor Johnston, Fall 2010). Strang (1991) suggests to study rate of change of the area under a curve for simple functions, such as piecewise linear and step functions, from the beginning of the course.

In order to help future teachers gain understanding of the inverse nature of the process of differentiation and integration, they go through the following activity, working in small groups.
For a continuous function f, define a new function F in the following way
$F(x) = \int_{a}^{x} f(t)dt$. Discuss in your group how this new function represents
the area under the graph of f.
For a small h, draw a picture of $F(x+h) - F(x)$. Compare this area increment to the area of the rectangle $hf(x)$. What does this tell you about $\dfrac{F(x+h) - F(x)}{h}$ compared to $f(x)$?
Teachers consider the area under the graph of a continuous function f from a fixed point a to x as a function of x, $F(x)$. The rate of change of the area

at x will be approximately the change of area from x to $x + h$ divided by the increment h, that is, $\dfrac{F(x+h)-F(x)}{h}$. If h is small, the increment of the area $F(x+h) - F(x)$ will be very close to the area of the rectangle $f(x)h$ (see Figure 44). For small h the difference between the increment of the area under the curve and the area of the rectangle is small compared with h. Therefore, $\dfrac{F(x+h)-F(x)}{h}$ will be very close to $\dfrac{f(x)h}{h}= f(x)$. That is, $\displaystyle\lim_{h \to 0}\dfrac{F(x+h)-F(x)}{h} = f(x)$. Thus, the instantaneous rate of change $F'(x)$ of the area under the curve at x is given by the value $f(x)$. Students will thus have a basis to see that the derivative of the integral of a function is the original function.

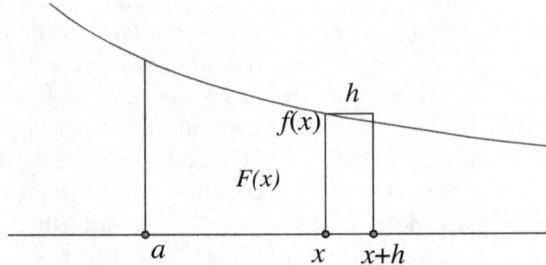

Figure 44. Change in the area under the curve.

Additional issues in the learning of calculus.

In addition to giving teachers the opportunity to develop connections and a deeper understanding, by experiencing themselves activities that they could in the future also use in their own students, the methods course also strives to give them the opportunity to reflect on and address some the issues students face when learning calculus.

Misconceptions about limit.
The first activity is geared to help teachers get a better understanding of students' misconceptions of the limit of a sequence and elicit from teachers strategies to prevent the formation of misconceptions or address them. An article on the difficulties students face when learning the concept of limit (Jacobs, 2002) is made available to the future teachers. They are asked to read in class the section on misconceptions about limits (p. 6-8). Jacobs lists four kinds of misconceptions.

    1. A limit is a number that can *never* be attained.

2. A limit is a number that is not to be exceeded.

3. Eventually the limit is reached.

4. A limit is evaluated by plugging into a formula.

In their small groups teachers choose one of the misconceptions about limit addressed in the article. They design an intervention to prevent or redress the formation of such misconception in calculus students. They outline their approach on paper with enough detail so their team can share with the rest of the class. One of the questions in the final exam is about what examples and non-examples they would use to introduce the concept of limit of a sequence in order to help students develop a more complete understanding of the concept. They also have to discuss what possible misconceptions about limit may be prevented or addressed by their examples and non-examples.

Developing advanced understanding of the derivative

An activity to help future teachers focuses on what Zandieh (2000) calls three layers in the learning of derivative. The first layer is the ratio layer, the second the limit layer, and the third is the function layer. Of course these three steps in the teaching of derivative are not new. The first two correspond to what Kline (1967) calls the method of increments. However, the work of Zandieh shows that students need to understand well each of these layers, one by one, to develop an advanced understanding. Future teachers are given a handout based on a chapter published by Sealy and Flores (2002) that discusses the three layers. An excerpt is presented here. In the handout given to teachers, figures similar to those used elsewhere in this chapter are included too (Figures 16 and 17). To avoid repetition they are left out in this section.

Layer one: The ratio of increments. In the layer of the ratio or quotient of increments, we can see the formula $\dfrac{f(x_0 + h) - f(x_0)}{h}$ as the slope of a secant to the graph of a function f between two points x_0 and $x_0 + h$. In the case of a function $d(t)$ that represents the distance traveled by an object with varying speed, the quotient $\dfrac{d(t_1) - d(t_0)}{t_1 - t_0}$ represents the average speed between times t_0 and t_1. Students in a calculus class should be very familiar with the symbolic notation of the slope of a straight line as: $\dfrac{y_2 - y_1}{x_2 - x_1}$ or $\dfrac{\Delta y}{\Delta x}$.

The quotient of increments corresponds to the quotient of the opposite leg over the adjacent leg in the triangle of Figure 45, that is, the tangent of angle α.

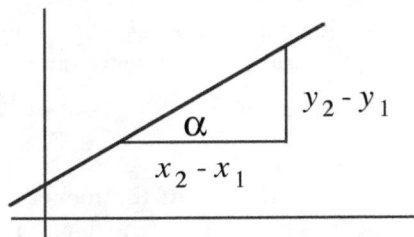

Figure 45. Slope of a line.

Our first step is to show students that the slope of the secant is the slope of the corresponding straight line between two points. At first, usually students interpret $\frac{y_2 - y_1}{x_2 - x_1}$ as the *process* of dividing one number by another.

Next, we would like for our students to recognize this ratio as one specific number (or as an *object*), which is the result of the process of dividing one difference by another. The words *process* and *object* derive from the work of Sfard (1992).

Seeing a ratio (or the slope of the secant) as an object is desirable before moving on to another layer of the derivative. Zandieh (2000) refers to this view of a ratio as a "process-object pair". Students should be able to understand both the process of dividing two quantities and the resulting object. When a student can move from the process to the object and then back to the process (when necessary), the student has constructed an appropriate "process-object" pair.

Layer two: The limit. In the limit layer, the expression $\lim_{h \to 0} \dfrac{f(x_0 + h) - f(x_0)}{h}$, is often thought of as the slope of the tangent of a function at a particular point x_0. We want to emphasize that we are talking about x_0 as a particular value, and not as a variable. When we add the function layer, we will examine the slope of the tangent at *every* point, thus referring to x as a variable.

Because the ratio layer should already be fairly well established with students, we will use it as a foundation for the limit layer. For this layer, we will again look at slopes of secant lines from x to $x_0 + h$, but now we will let h get successively smaller. The limit of the values of the slopes of the secants will be the slope of the tangent to the curve at x_0.

Layer three: The function. After building layers one and two (the ratio and the limit), students should be able to calculate the derivative of a function at any point x_0. They should be able to do this graphically using the slope of the tangent, and symbolically using the difference quotient. They should be familiar with the symbolic representation $f'(x_0) = \lim_{h \to 0} \dfrac{f(x_0 + h) - f(x_0)}{h}$, and understand that this representation

allows them to compute the derivative of the function at any point (or at least any point where the function is differentiable). The ultimate goal for this layer is to combine all of these points in to an entire function $f'(x) = \lim_{h \to 0} \dfrac{f(x + h) - f(x)}{h}$.

Teachers read the handout on the three layers and then were asked to design activities for students to better understand each layer, and help them move from one layer to the next. After the future teachers went through the activity one of them wrote in her reflection about the fact that students are rushed and often not given the opportunity to think through the first two layers. "Some students are taking calculus courses, but the textbooks and teachers are skipping over parts to help students fully understand the meaning of the derivative. They jump to the 3rd layer of the function before addressing it as a ratio or limit. I have experienced this both times I took calculus in high school and in college that my teachers jumped right to the function layer." (Marie Young, Fall 2010)

Concluding remarks

Skemp (1987) has pointed out *"the great psychological difficulty for teachers of reconstructing their existing and longstanding schemas"* (p. 161). Teachers who learned calculus (and most of their mathematics for that matter) focusing on procedural proficiency rather than conceptual understanding face such difficulties. A few sessions or even one semester is certainly not enough time for teachers to reconstruct their schemas, but at least they can be made aware that such a change is needed. The students' quotes included in this chapter are indication that some of them are aware that the way they or their fellow students learned calculus could be greatly improved by making the underlying concepts more explicit. As the new teachers begin their teaching careers, they will need the will and the time to continue to develop their knowledge and resources on how to emphasize conceptual understanding in their teaching of calculus and other topics in mathematics.

References
Barnes, J. (2011). Feather boas in real analysis. *PRIMUS, 21*(2), 130-141.

Brenes Castro, V., Díaz Campos, M. A., & Gómez Solano, J. (1981). *La enseñanza de la geometría*. San José, Costa Rica: Universidad de Costa Rica.

College Board. (2010). AP Examination volume changes (2000-2010). Retrieved June 6, 2011 from http://professionals.collegeboard.com/data-reports-research/ap/data

College Board. (2011). Sample questions & scoring guidelines. Retrieved June 10, 2011 from http://www.collegeboard.com/student/testing/ap/calculus_ab/samp.html?calcab

Cornu, B. (1991). Limits. In D. Tall (Ed.), *Advanced mathematical Thinking* (p.153-166). Boston: Kluwer.

Cory, B. L. (2009). Visualizing the limits of sequences [Electronic Version]. *On-Math: Online Journal of School Mathematics* Retrieved June 6, 2011 from http://www.nctm.org/resources/view_article.asp?article_id=8937&page=1.

Cory, B. L. (2010). Bouncing balls and graphing derivatives. *Mathematics Teacher, 104*(3), 206-213.

Cory, B. L., & Garofalo, J. (2011). Using dynamic sketches to enhance preservice secondary mathematics teachers' understanding of limits of sequences. *Journal for Research in Mathematics Education, 42*(1), 65-96.

Cory, B. L. & Smith, K. W. (in press). Delving into limits of sequences. *Mathematics Teacher.*

Cottrill, J., Dubinsky, E., Nichols, D., Schwingendorf, K., Thomas, K., & Vidakovic, D. (1996). Understanding the limit concept: Beginning with a coordinated process scheme. *Journal of Mathematical Behavior, 15*(2), 167-192.

Courant, R. and Robbins, H. (1941). *What is mathematics?* NY: Oxford University Press.

Courant, R. and Robbins, H. (1996). *What is mathematics?* (2nd edition, revised by Ian Stewart). NY: Oxford University Press.

Cummins, K. (1977). A student experience-discovery approach to the teaching of calculus. In L. S. Grinstein & B. Michaels (Eds.), *Calculus: Readings from the Mathematics Teacher* (p. 31-39). Reston, VA: National Council of Teachers of Mathematics.

Davis, R. B., & Vinner, S. (1986). The notion of limit: Some seemingly unavoidable misconception stages. *Journal of Mathematical Behavior, 5*, 281-303.

Flores, A. (1991). Calculators in calculus: That's the limit. *PRIMUS, 1*(3), 295 - 301.

Flores, A. (1992). Geometric sequences, squaring, and square roots. Three approaches to the golden ratio. In J. T. Fey (Ed.), *Calculators in Mathematics Education, 1992 Yearbook* (p. 241-244). Reston, VA: National Council of Teachers of Mathematics.

Flores, A. (2002). If *pi* were equal to 3.... *Ohio Journal of School Mathematics*, *46*, 41-44.

Flores, A., & McLeod, D. (1990). Calculus for middle school teachers using computers and graphing calculators. In *Proceedings Third Annual Conference on Technology in Collegiate Mathematics*. Columbus, OH: The Ohio State University.

Flores, A. & McLeod, D. (1993). Cálculo con computadoras y calculadoras gráficas. *Cuadernos de Investigación*, No. 23, Año 7 (Feb), 41-53. Departamento de Matemática Educativa CINVESTAV.

Flores, L. & Flores, A. (2005). Cálculo, papel, tijeras. *Eureka, 20*, 34-38.

Flores, L. & Flores, A. (2006). Calculus, paper, scissors. *PRIMUS, 16*(4), 358-362.

Flores Peñafiel, A. (1990). Un límite interesante con una tira de papel. *Educación Matemática, 2*(2), 61-63.

Flores Peñafiel, A. (1991). Las calculadoras en cálculo: El límite. *Educación Matemática, 3*(1), 110 -119.

Fremont, H. (1978). *Teaching secondary mathematics through applications* (2nd ed.). Boston: Prindle, Weber, & Schmidt.

Goldstine, H. H. (1972). *A history of numerical analysis from the 16th through the 19th centuries*. NY: Springer.

Hitt, F. & Páez, R. (2001). The notion of limit and learning problems. In R. Speiser, C. A. Maher, C. N. Walter (Eds.), *Proceedings of the twenty-third annual meeting North American Chapel of the International Group for the Psychology of Mathematics Education* Vol. 1 (p. 169-176). Columbus, OH: ERIC Clearing House for Science, Mathematics and Environmental Education.

Hughes-Hallett, D., Gleason, A. M., McCallum, W. G., et al. (2009). *Calculus: Single and multivariable* (5th ed.). New York: John Wiley.

Husch, L. S. (2001). Fundamental theorem of calculus. Retrieved June 10, 2011 from http://archives.math.utk.edu/visual.calculus/4/ftc.2/

Infeld, L. (1978). *Whom the gods love: The story of Evariste Galois*. Reston, VA: National Council of Teachers of Mathematics.

Jacobs, S. (2002). Why is the limit concept so difficult for students? *The AMATYC Review, 24*(1), 25-34.

Klein, F. (2004). *Elementary mathematics from an advanced standpoint: Arithmetic, algebra, analysis*. NY: Dover, 2004.

Kline, M. (1967). *Calculus: An intuitive and physical approach*. New York: Wiley.

Kline, M. (1998). *Calculus: An intuitive and physical approach* (2nd edition). New York: Dover Publications.

Nelsen, R. B. (2000). *Proofs without words II*. Washington DC: Mathematical Association of America.

Nitecki, Z. (2009). *Calculus deconstructed: A second course in first-year calculus*. Washington, DC: Mathematical Association of America.

Olson, Alton T. (1977). *Mathematics through paper folding*. Reston, VA: National Council of Teachers of Mathematics.

Pacific Tech (2007). *Graphing Calculator* 4.0 [Computer software]. Berkeley, CA: Pacific Tech.

Roh, K. (2010). How to help students conceptualize the rigorous definition of the limit of a sequence. PRIMUS, *20(6)*, 473-487.

Sawyer, W. W. (1961). *What is calculus about?* Washington, DC: Mathematical Association of America.

Sealey, V. & Flores Peñafiel, A. (2005). Entender la derivada: Sí se puede. In F. Hitt & J. C. Cortés (Eds.), *Reflexiones sobre el aprendizaje del cálculo y su enseñanza* (p. 175-196) Morelia, Michoacán: Editorial Morevallado.

Sfard, A. (1992). Operational origins of mathematical objects and the quandary of reification—The case of function. In G. Harel & E. Dubinsky (Eds.), *The concept of function: Aspects of epistemology and pedagogy* (p. 59-84). Washington, DC: Mathematical Association of America.

Skemp, R. R. (1987). *The psychology of learning mathematics*. Hillsdale, NJ: Lawrence Erlbaum Associates.

Strang, G. (1991). *Calculus*. Wellesley, MA: Wellesley-Cambridge Press.

Thurston, W. P. (1998). On proof and progress in mathematics. In T. Tymoczko (Ed.), *New directions in the philosophy of mathematics* (revised and expanded ed.) (pp. 337-355). Princeton, NJ: Princeton University Press.

Toeplitz, O. (2007). *The calculus: A genetic approach*. Chicago: University of Chicago Press.

Vinner, S. (1991). The role of definitions in the teaching and learning of mathematics. In D. Tall (Ed.), *Advanced mathematical thinking* (p. 65-81). Dordrecht, The Netherlands: Kluwer Academic Publishing.

Voellmy, E. (1974). *Jost Bürgi und die Logarithmen*. Basel: Birkhäuser.

Williams, S. R. (1991). Models of limit held by college calculus students. *Journal for Research in Mathematics Education, 22*(3), 219-236.

Zandieh, M. (2000). A theoretical framework for analyzing student understanding of the concept of derivative. In E. Dubinsky, A. Schoenfeld, & J. Kaput (Eds.), *Research in Collegiate Mathematics Education, IV.* (p. 103-112). Providence, RI: American Mathematical Society.

3 ORANGE YOU GLAD I DID SAY "FRACTION DIVISION"? [3]

In this article we address how teachers can deal with the issues and tensions created by trying to accomplish two important goals when teaching division of fractions. The first is to let students make sense of division of fractions on their own, by working individually and in small groups, using concrete or pictorial representations, inventing their own processes, and presenting and justifying their answers and processes to each other. The second is to help students develop their understandings of division of fractions to a deeper level, as a multiplicative comparison, so that they develop their proportional thinking. When comparing two quantities or numbers multiplicatively a ratio is involved. To help students develop their ability to compare fractions multiplicatively we need to encourage them to think beyond iterative strategies such as repeated addition or repeated subtraction. The teacher may have to intervene at crucial points to redirect the thinking of the students to focus on multiplicative comparisons of fractions.

Seventh-graders making sense of division of fractions on their own
In this section we illustrate ways in which four middle school students, solving division of fraction problems in the context of servings, made sense of division of fractions on their own. The examples come from a larger study by the second author (Day 2010). The purpose of her study was to understand how students justified their strategy and solution to their peers,

---

[3] Flores, A. and Priewe, M. (2013). Orange you glad I did say "fraction division"? *Mathematics Teaching in the Middle School, 19*(5), 288-293. Copyright National Council of Teachers of Mathematics. Used by permission.

so oftentimes the teacher/researcher listened carefully but did not intervene. The teacher and students agreed that showing your mathematical thinking includes more than a simple list of steps or an algorithm. Students communicated their thoughts through graphic representations, symbols, and words. To justify their reasoning students applied three guiding principles: convince themselves by making personally-meaningful solutions to problems; convince others by communicating their understanding through graphic representations, words, and symbols; and make sense of other students' justifications to raise challenges if disagreements occurred.

Story problems in the context of finding out how many servings can be formed with a given quantity, when the size of the serving is known, lend themselves to the use of measurement interpretation of division, and the use of repeated subtraction to find the answer. This setting has the advantage of allowing students to use concrete or pictorial representations and find their own strategies (Kribs-Zaleta 2008).

When using measurement interpretation of division, the simplest case is when the size of the serving fits exactly into the amount being distributed. For example, if a serving is ¾ of an orange, and there are 4½ oranges, there will be exactly 6 servings. A more difficult situation is when the size of the serving does not fit exactly a whole number of times into the amount, and there is a fractional remainder. Usually students working on problems with a fractional remainder struggle to make sense of the remainder in terms of the unit used to measure (serving), rather than in terms of the objects used (oranges or cupcakes). We focus first on the interactions of four middle school girls—Taylor, April, Rosalee, and Rebecca (pseudonyms)—as they worked on the following problem:

> A serving is ¾ of an orange. There are two and a half oranges. How many servings (including parts of a serving) can be made?

Taylor, April, Rosalee, and Rebecca used concrete objects and pictures to represent the problem and its solution (Figure 1).

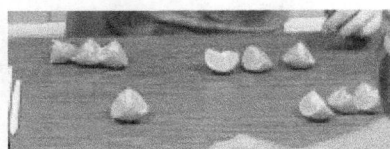

Figure 1. April showing each student serving and the part remaining

In communicating their thinking to one another, there was confusion because the students did not explicitly define their "whole" (an orange or a serving). This caused disagreement in the final solution to the problem.

Taylor: It says. Suppose you have two and a half oranges. One…two…three. I just did three because I thought it would be easier for me. So this is one serving [She shaded in three-fourths of an orange.] and this is another [She shaded in another three-fourths of an orange.]…this is another serving [She shaded in three-fourths of an orange.]…and yea. So this is technically not there [Taylor pointed to the last half of the third orange.]. You're not supposed to…so there's three servings right there. There's one left over. And there's out of three possible…So you know how you need three like three-fourths…

Rosalee: Yea.

Taylor: …to get one serving. So that's like seventy-five cents, but instead of like doing that you have one out of three possible. So I put three and one-third. 'Cause it's one-third left over.

The unit specified in the question is that of a serving. Students correctly figured out how many complete servings were there, by using repeated subtraction of servings, but struggled with interpreting the remainder. The remainder is 1/4 of an orange, but 1/3 of a serving.

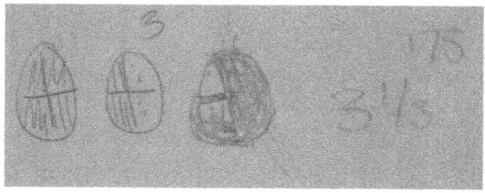

Figure 2. Taylor's graphic representation

Figure 3. Taylor showing 1/3 of a serving.

Taylor used her graphic representation (Figure 2) to show the remaining amount on the measurement division fraction problem, but was not always explicit when using her symbols and words what the remaining part represented of a whole unit (such as servings). Therefore, at times she struggled to communicate effectively with others when trying to explain her strategy.

Taylor's solution of three and one-third was correct, but she did not specify that it was three and one-third servings. Taylor said that there was one left over. Based on a preliminary written survey and her oral interview with the teacher it was evident that Taylor understood that the remaining amount represented one-fourth of an orange. However, she was not explicit that the remaining amount represented one-fourth of an orange when she explained it to her group.

Later in the discussion, Taylor continued to try to convince her group by being explicit that the remainder was not one-fourth by saying, "But it's not a fourth of a serving" (Figure 3). The girls had mentioned servings before, but this was the first time they discussed a fraction of a serving. It may have helped her group to understand her if Taylor would have said that it was one-fourth of an orange and then tried to explain that it took three one-fourths of an orange to make a full serving.

The students were using graphic representations to show the remaining amount as well as symbols and words. However, they needed to be more explicit when using their symbols and words to explain what the remaining amount represented in units as a whole, so they could communicate when they were referring to different wholes.
Later in an interview with the teacher, Taylor stated that she needed to explain her answer more clearly.

Taylor: Like the problem said you were supposed to look at the serving and not the oranges. Well, I never said that in my answer. And I think if I would have explained that better they would have come to know like how I got the answer.

Although Taylor and her group realized that they needed to be explicit about the unit, this error of not making the unit explicit was persistent throughout the study. The following sections of this article suggest ways in which teachers can help students step back so that they can develop their mathematical understandings of division of fractions at a deeper level.

Connecting division of whole numbers and fractions
By stepping back to consider division problems with whole numbers and interpreting the result as a fraction, students establish an important connection that is not always present. Some children will notice on their own that the same numbers appear in a division problem like $7 \div 4$ and in the answer, 7/4. On their own, or with teacher's guidance, students need to realize that the fact that the same numbers appear in the original division

problem and in the answer expressed as a fraction $7 \div 4 = 7/4$ is not a coincidence. The realization that each person's portion in a distribution situation can be predicted by the (multiplicative) relation between objects and participants is one example of the close relation between proportions and fractions and how the concepts are intertwined (Streefland, 1991, p. 130). This realization will be a stepping stone later for thinking in terms of ratio when dividing fractions.

One way teachers may help students understand the remainder in a division of fractions problem is to direct their attention to similar problems using whole numbers. For example, consider the following problem.

A serving is 3 cupcakes. There are 7 cupcakes. How many servings are there?

One answer is 2 servings, and one cupcake as remainder. Students need to interpret the remainder in terms of the serving. Students need to realize that they need to divide the remainder by the size of the serving, that is, that the remainder too needs to be divided by the divisor. If a serving is 3 cupcakes, then one cupcake is $1/3$ of a serving. Therefore, another answer is 2 and $1/3$ servings. When students write the answer to $7 \div 3$ as 2 R1, they need to realize that the 2 refers to how many servings, not cupcakes, but the remainder 1 refers to cupcakes.

A teachable moment: Dealing with the remainder as a quotient
While Taylor and April were working on the following problem they made explicit for the first time in the study a way to compare the remainder with the serving multiplicatively.

Adam has been serving $2/3$ cup of lemonade to each student. If he has $1\frac{1}{2}$ cups of lemonade left, how many students can still get lemonade? How much of a serving will the last student get?

To solve the problem, both girls added $2/3 + 2/3 = 1 + 1/3$, Taylor in her head, and April using a picture. Then they found the difference to $1\frac{1}{2}$. They both thought the answer was 2 student servings and $1/6$ of a serving. However, Taylor at one point said $1/6$ **of a cup** and then a minute or so later she said $1/6$ **of a serving**. She changed what she said without even realizing it. When the teacher reminded them about the orange problem they came to the conclusion that they were mixing up the units. They realized that the 2 represented student servings while the $1/6$ represented cups of lemonade.

Taylor: That would be cups of lemonade left over. And then we have to figure out if it takes…if it takes two-thirds and you have one-sixth left over. So if you had one-sixth and it needs to go into two-thirds of a cup of lemonade, how would we do that?

Taylor's statement, "That would be cups of lemonade left over," is an indication that she realized that the remaining 1/6 represented a fraction of a cup, not a 1/6 of a serving. April suggested how to deal with the remainder to express it as part of a serving.

April: One-sixth divided by two-thirds?

Thus the girls contemplated the idea of dividing the remainder by the serving, but could not quite remember the procedure for how to divide fractions, so they could not follow this approach. This is one example where lack of computational skill can prevent further conceptual exploration. There are several approaches a teacher can take in situations like this. One is for the teacher to make a note of the situation and ask students to place their thoughts on hold, and then help them remember or relearn how to divide fractions, so that they can continue with the problem. One way to divide fractions that highlights the connection of ratio with division is using common denominator. Their problem $1/6 \div 2/3$ is equivalent to $1/6 \div 4/6$. In this division problem we need to compare, by forming a ratio, the number of pieces in the group that represents 1/6 with the number of pieces in the group that represent 4/6. Because the pieces in each group are of the same size, the result will be equal to the ratio of the number of pieces in each group, that is, the ratio of the numerators, $1 \div 4$, or 1/4. Another approach is to take advantage of the calculating technology. Nowadays fortunately there are inexpensive fraction calculators available, so that a teacher can encourage students to use such tools to explore their ideas. The context, with some teacher's guidance, will help students interpret the result given by the calculator. Later, the teacher can come back and make sure that students know and understand how to divide fractions even when they do not have access to a calculator, but for the time being not remembering the procedure should not deter students from developing their conceptual understanding. With the fraction calculator, using the division key $\div$, the girls could have obtained $1/6 \div 2/3 = 1/4$. The important thing at this point is for teachers to help students to continue thinking about the comparison of two fractions (the remainder and the serving in this case) in multiplicative terms.

Reconnecting repeated subtraction to division
The initial approach used by Taylor and her teammates was to use the serving as a measurement unit and do repeated subtraction or repeated addition to figure out how many servings could be formed. Students need to realize that repeated subtraction can also be done in one step by using

division. Students can refer first to a familiar situation with whole numbers, such as how many pairs can be formed with six students. Students can easily see that the problem can be solved by repeatedly subtracting 2 from 6, or by dividing $6 \div 2$. Students can then go back to the situation with fractions and divide $2\frac{1}{2} \div \frac{3}{4}$ with a fraction calculator using the $\div$ key. After simplifying, the calculator will display the answer as 3 1/3, the same they obtained using their own method. Students need to remember the 1/3 refers to servings. Some fraction calculators also have the feature of division with remainder. For example, using the $\div$R key on a Casio Fraction Mate calculator, students would obtain the answer 3 R 0.25. Students can recognize as $0.25 = \frac{1}{4}$, and realize that this remainder refers to oranges.

In a discussion of the power of mathematical notation, Hiebert and Behr (1988) point out that mathematical symbolism can take students beyond their level of conceptual understanding. With calculators it is even more the case that students may easily find answers that they do not understand. Therefore we need to take the time to make sure students understand why the different operations and the corresponding keystrokes were chosen and what the answer means. In the example above, students need to relate the two different answers 3 1/3 and 3 R $\frac{1}{4}$ and make explicit what each of the numbers, 3, 1/3, $\frac{1}{4}$ represent in terms of servings and oranges.

To deepen students' understanding of the remainder as part of an orange or as fraction of a serving, the teacher may point to the relation of division with multiplication. In order to get the original dividend from the answer by using the divisor as a factor, students need to realize that in one case they would write $(3\ 1/3) \times \frac{3}{4} = 2\ \frac{1}{2}$, whereas when using the remainder, they would need to write $3 \times \frac{3}{4} + \frac{1}{4} = 2\ \frac{1}{2}$.

Concluding remarks

To develop their proportional thinking students need to shift from the use of composed-unit strategies such as iterating and partitioning to multiplicative comparisons (Lobato & Ellis 2010, p. 69). As illustrated above, this transition is not easy for students, and teachers need to be alert to recognize situation where asking students the right question, providing them with an appropriate tool, or pointing to a similar connection can help students make the transition from additive comparisons to multiplicative comparisons in the context of division of fractions.

References

Day, Melina M. *Middle school mathematics students' justification schemes for dividing fractions.* Unpublished Doctoral dissertation. Tempe, AZ: Arizona State University, 2010.

Hiebert, James, and Behr, Merlyn (Eds.). *Number concepts and operations in the middle grades.* Reston, VA: National Council of Teachers of Mathematics, 1988.

Kribs-Zaleta, C. Oranges, posters, ribbons, & lemonade: Concrete computational strategies for dividing fractions. *Mathematics Teaching in the Middle School* 13(2008): 453-457.

Lobato, Joanne, and Ellis, Amy B. *Developing essential understanding of ratios, proportions and proportional reasoning.* Reston, VA: National Council of Teachers of Mathematics, 2010.

Streefland, Leen. *Fractions in realistic mathematics education: A paradigm of developmental research.* Dordrecht, The Netherlands: Kluwer Academic Publishers, 1991.

4 VISUALIZING THE STATES[4]

In this article, participants estimate the areas of other states in relation to North Carolina three times. The resources available to participants increase each time. First they use only their individual mental images of the sizes of the states, second they compare their estimates with that of other participants and explain their reasoning behind their estimates, and finally, they have access to a map of the states.

Mental maps. People constantly create and keep mental images. People carry mental pictures and maps in their minds. Why do you think people need these mental maps? Can you picture a map of the United States in your mind? Picture the location and size of different states, such as North Carolina, Virginia, South Carolina, Arizona, Alaska, California, Texas, Delaware, Florida. Would you like to know how accurate your picture is? Visualize how big is North Carolina in relationship to other states. How does it compare to Virginia or California? How about Alaska?

First estimation (individual).

Let's use the area of North Carolina as the unit of measurement. Write in the first column of Table 1 what you think the ratios are for the sizes of the other states compared with North Carolina, that is, the ratio area of other state / area of North Carolina. If the other state is bigger than North Carolina, the ratio will be bigger than 1. If the other state is smaller, the ratio will be smaller than 1. In other words, estimate how many North Carolinas will fit into each of the other states on the list.

---

[4] Flores, A. (2013). Visualizing the states. *The Centroid, 39*(2), 6-9. Reprinted by permission North Carolina Council of Mathematics.

Second estimation. Discussion and possible revision.

Compare and discuss your estimate with your neighbors. Explain the reasoning behind your estimate. If you wish to change your estimate, write a new estimate in column two.

Table 1. Estimation of ratios of areas

State	Guess 1	Guess 2	Guess 3	Actual area	Ratio
North Carolina	1				1
Virginia					
South Carolina					
Georgia					
Arizona					
Alaska					
California					
Texas					
Delaware					
Florida					

Third estimation. Look at a map.

Now look at a map with all 48 contiguous states marked on it (Figure 1). How does it compare with your mental image? Do you still think your second estimate is accurate? When you see the actual map it makes you wonder how your mental image could be so far off. If you want to revise your estimates, write them in column 3.

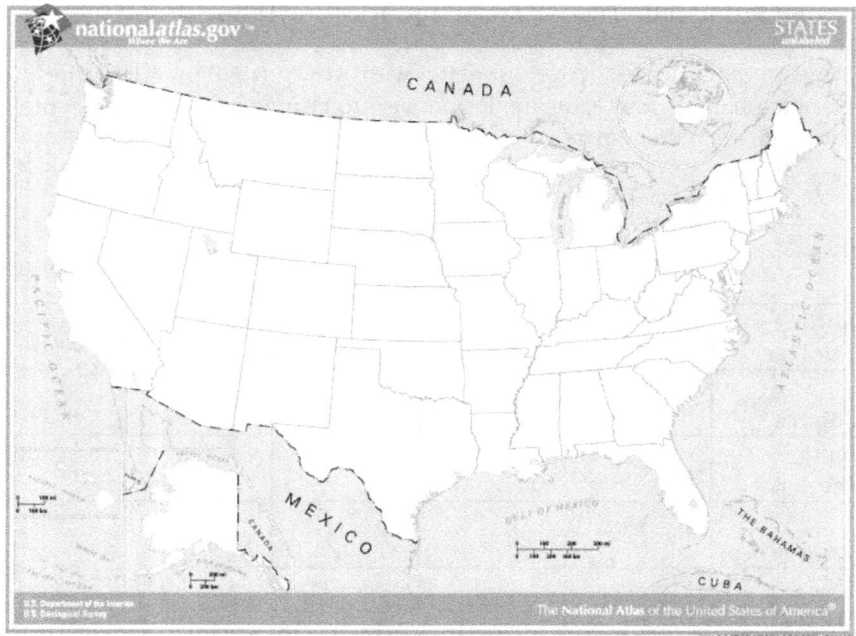

Figure 1. 48 contiguous states with Alaska and Hawaii at different scales
Map obtained from the National Atlas of the United States, which is in the
public domain. Available: http://en.wikipedia.org/wiki/File:National-atlas-
blank-state-outlines.png

Notice that the map, like many school maps, displays Alaska and Hawaii in
the lower left corner of the map and with a different scale, making Alaska
look much smaller than it really is. Even when the difference in scale is
made explicit, it is very hard to form a mental image of how big Alaska
really is. If we would display Alaska at the same scale, we would see how big
it is compared to the 48 contiguous states (Figure 2). When comparing
Alaska to Texas, people often do not realize that the area of Alaska is more
than twice the area of Texas. Or as one Alaskan explained to a Texan, if you
would cut Alaska in half to make two states, Texas would be the *third*
biggest state.

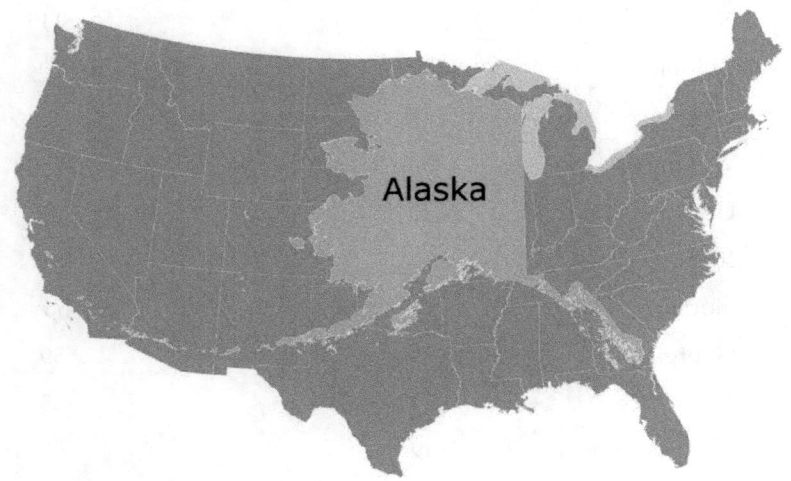

Figure 2. Alaska compared to the 48 contiguous states.
Permission to share under the terms of the GNU Free Documentation
License. Available http://en.wikipedia.org/wiki/File:Alaska-Size.png

Using the actual areas to compute the ratios

At the end, students can use the actual area values given in Table 2 to estimate the ratios of areas of other states to area of North Carolina. The value for each state, taken from the Census Bureau, represents the total area which includes land area and water area. The water area includes inland, coastal, Great Lakes, and territorial waters. The remark above about the relative size of Alaska and Texas still holds if we consider only land area, 572 for Alaska and 262 for Texas (in thousands of square miles). Other sources report total areas for states in different ways, for example, including inland waters, but excluding coastal waters.

Table 2. Total areas of states

Rank	State	Area (thousands of km²)	Area (thousands of square miles)
1	Alaska	1718	663
2	Texas	696	269
3	California	424	164
4	Montana	381	147
5	New Mexico	315	122
6	Arizona	295	114

7	Nevada	286	111
8	Colorado	270	104
9	Oregon	255	98
10	Wyoming	253	98
11	Utah	220	85
13	Idaho	216	84
22	Florida	170	66
24	Georgia	154	59
27	New York	141	55
28	North Carolina	139	54
29	Arkansas	138	53
35	Virginia	111	43
40	South Carolina	83	32
44	Massachusetts	27	11
49	Delaware	6.4	2.5
50	Rhode Island	4.0	1.5

Concluding remarks

We adapted this activity from one about Australia's states (Lovitt and Clarke 1988). The activity can be adapted in turn for other area comparisons, such as comparing the area of the school building to the school stadium, to the total area of the school, or to the areas of houses or other facilities or buildings in the city.

References

List of U.S. states and territories by area. Available: http://en.wikipedia.org/wiki/List_of_U.S._states_and_territories_by _area

Lovitt, C. and Clarke, D. Map of Australia. In *Activity Bank*, vol. 2 (pp. 317-322). The Mathematics Curriculum and Teaching Program, 1988.

U.S. Census Bureau, 2000 Census of Population and Housing, *Population and Housing Unit Counts* PHC-3-1, United States Summary (Table 17), Washington, DC, 2004. *Available: www.census.gov/prod/cen2000/phc3-us-pt1.pdf.*

5 DEVELOPING THE ART OF SEEING THE EASY WHEN SOLVING PROBLEMS[5]

For Leonardo da Vinci "saper vedere", that is, knowing how to see, or having the art to see, was the key to unlock the secrets of the visible world. *Saper vedere* included precise sensory intuitive faculty as well as artistic imagination (Heydenreich 1954) which are at the root of his inventiveness and creativity. According to Leonardo, to understand, you only have *to see things properly* (Bramly 1994, p. 264). Knowing how to see is also important in mathematics. The Italian mathematician Bruno de Finetti (1967) wrote a book on "Saper vedere" in mathematics. He highlights several aspects of knowing how to see in mathematics, such as knowing how to see the easy, how to see the concrete things, and how to see the economic aspects. He also discusses in what ways knowing how to see helps us also recognize better the meaning of the general and systematic methods of mathematics that are represented in formulas. De Finetti starts his book highlighting the importance of reflection for learning the art to see. Reflection also plays a central role in Polya's *Looking back* stage in problem solving.

In this article we will focus on learning the art of seeing the easy, using an example of a problem posed to future secondary mathematics teachers. De Finetti points out that it is often difficult to see the easy things, that is, be able to distinguish, in the complexity of circumstances present in a problem, those that are enough to formulate the problem or that allow to do the

---

[5] Flores, A. and Braker, J. (2013). Developing the art of seeing the easy when solving problems. *Mathematics Enthusiast Journal*, *10*(1 & 2), 365-378. Reprinted by permission.

formulation as several successive steps that can be carried out easily.

In this article we will first present the strategy used by the future teachers, and then an approach gained by looking back at the problem and trying to see it at a glance. We finish with a brief discussion of why it is worthwhile for prospective teachers to look back at the problem.

The problem

During a course for prospective high school teachers, one of the assignments was to present a problem for their fellow students that could be modeled or solved with high school mathematics. The second author posed the following problem to her classmates.

> You are attempting to bathe a cat in your kitchen. Unfortunately, the cat is not as open to the bath as you were hoping, and as a result you spill 3 gallons of water in your kitchen. Which brand of paper towel should you use to clean up the spill?

Brand A	Brand B
Paper towel is 1/32 inches thick	Paper towel is 1/64 inches thick
Total diameter of roll is 5 inches	Total diameter of roll is 6 inches
Diameter of hollow inside is 2 inches	Diameter of hollow inside is 2 inches
One sheet absorbs 1.5 fluid ounce	One sheet absorbs 1 fluid ounce
Each sheet is 10 inches long	Each sheet is 10 inches long

The assumption is that the price for the roll is the same for both brands. Remember that 1 gallon = 128 fluid ounces.

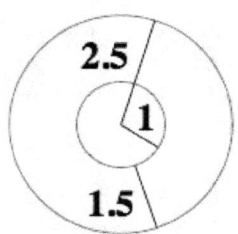

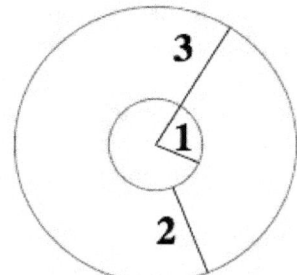

Figure 1a. Cross section Brand A roll

Figure 1b. Cross section Brand B roll

The approach used by all the future secondary teachers was to find how many rolls of each brand were needed to clean up the spill. To find this number they decided to compute how much water can be absorbed by one roll of each brand, finding first how many sheets are in each roll. The future teachers modeled the spiral cross section of the role of paper as a series of concentric circles. Each successive layer was a little longer because the thickness of each sheet increased the diameter.

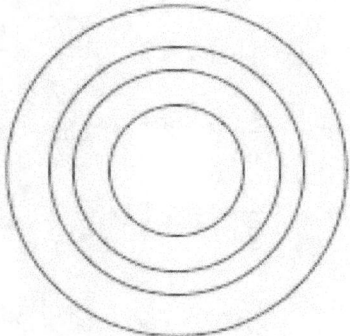

Figure 2. Concentric layers

Thus for Brand A the first layer has a length of $C_1 = 2\pi \times 1$

For the second and third layers the length is $C_2 = 2\pi\left(1+\dfrac{1}{32}\right)$,

$C_3 = 2\pi\left(1+\dfrac{2}{32}\right)$ and in general, the length of the k-th layer is

$C_k = 2\pi\left(1+\dfrac{k-1}{32}\right)$.

The number of layers is given by dividing the thickness of the roll by the thickness of each sheet $n = \dfrac{1.5}{1/32} = 48$.

The total length is thus

$$C_1 + C_2 + \ldots C_{128} = 2\pi + 2\pi\left(1+\frac{1}{32}\right) + 2\pi\left(1+\frac{2}{32}\right) + \ldots + 2\pi\left(1+\frac{47}{32}\right)$$

$$= 2\pi\left(48 + \frac{1}{32}(1+2+\ldots+47)\right) = 2\pi\left(48 + \frac{1}{32} \times \frac{47 \times 48}{2}\right)$$

$$= 96\pi + \frac{141}{2}\pi \approx 523.$$ Thus the total length of a roll of brand A is 523

inches. The length of each sheet is 10 inches, so there are about 52 sheets per roll. These sheets can absorb $52 \times 1.5 = 78$ ounces of water. Thus each roll absorbs 78 fl. oz. of water. To clean 3 gallons = 3×128 fl. oz. = 384 fl.

oz. we need $\dfrac{384 \text{ fl. oz}}{78 \text{ fl. oz/roll}} = 4.9$ rolls. That is, we need almost 5 rolls of Brand A to clean the spilled water.

For Brand B the length of each layer is $C_k = 2\pi\left(1 + \dfrac{k-1}{64}\right)$ and the number of layers is $n = \dfrac{2}{1/64} = 128$. The total length is

$$C_1 + C_2 + \ldots C_{128} = 2\pi + 2\pi\left(1 + \dfrac{1}{64}\right) + 2\pi\left(1 + \dfrac{2}{64}\right) + \ldots + 2\pi\left(1 + \dfrac{127}{64}\right)$$

$$= 2\pi\left(128 + \dfrac{1}{64} \times \dfrac{127 \times 128}{2}\right) \approx 1602.$$ The total length is thus about 1602 inches. Because each sheet is 10 inches long, that is about 160 sheets. Each sheet absorbs one fluid ounce, so one roll absorbs 160 fl. oz. To clean 3 gallons we need $\dfrac{384 \text{ fl. oz}}{160 \text{ fl. oz/roll}} = 2.4$ rolls. Brand B is clearly the better choice for this problem.

Looking back

Polya points out that when we have obtained a long and involved solution we naturally want to see whether there is a more direct and clear way to solve the problem. He advises to question *Can you derive the result differently? Can you see it at a glance?* (Polya 1973, p. 61). He also points out that even when we have found a satisfactory solution we may still benefit from finding a different solution, which may give us further understanding or allow us to look at the problem from a different perspective. Polya encourages us to study the result and try to understand it better, to see a new aspect of it (p. 64). In the same way that we might get a better perception of an object by using two senses, we might get a better understanding of a problem by using two proofs. Future teachers need to learn to guide their students on how to find in the result itself indications of a simpler solution.

The approach used by the future teachers described above has several advantages. One is to highlight the use of an arithmetic sequence and how the average of the terms is used to obtain the sum. One way to read the formula for $1 + 2 + \ldots + n = n(n+1)/2$ is that we are multiplying the average of the terms, $(n+1)/2$ by the number of terms n. Another advantage is that we actually find how many rolls of paper we need.

In terms of the original problem posed we may want to look back and ask ourselves what are the essential differences between the two types of paper rolls for this problem. In the situation described, we really want to compare the efficiency to absorb water of the rolls relative to each other to determine which brand to use. Once we determine what brand to use then we can compute how many rolls of that brand we need.

Use of proportional reasoning to compare paper towels
A key insight for solving this problem in a different way is to realize that when comparing the rolls, we need to compare their ratios with respect to different factors that affect their number of sheets and absorption capacity.
In the solution above, the average of the lengths of the layers played an important role. Here we will see how we can use the average in a different way. The number of sheets in a roll will be proportional to the area of the circular ring cross section. The area of this ring can be obtained by multiplying the circumference of the average circle by the width of the ring (Figure 3). If r_1 is the radius of the hollow circular center, r_2 the radius of the paper roll, and and their respective diameters, then the area of the cross section is given by $A = \frac{1}{2}(d_1 + d_2)\pi(r_2 - r_1)$ (1)

Exercise 1. Derive formula (1) for the area of the ring using the difference of areas of concentric circles.

Exercise 2. Discuss in what ways formula (1) is analogous to the formula for the area of a trapezoid.

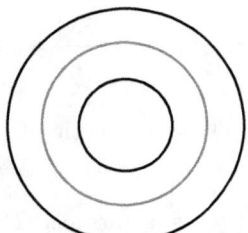

Figure 3. The average circle

Thus, a good way to compute how many sheets are around the roll is by using the average circle, in this case the mid circle between the hollow core and the outer layer. For Brand A this average circle has a diameter of $(2+5)/2 = 3.5$ inches, for Brand B a diameter of $(2 + 6)/2 = 4$ inches. Given that the lengths of the concentric layers (Figure 2) form an arithmetic progression, it is not surprising that we use the average. The number of sheets around will be proportional to the diameters of the mid circles, and proportional to the useful width of the rolls. The corresponding ratios comparing Brand B to Brand A will be thus 4/3.5 for the diameters of the circles, and 2/1.5 for the widths. The number of sheets will be inversely proportional to the thickness of each sheet, so the ratio between Brand B and Brand A is 64/32 = 2. Both brands have the same lengths of sheets, so to get the total ratio of sheets we just need to multiply these ratios. So Brand B has 4/3.5 × 2 × 2/1.5 more sheets than Brand A. Because the ratio of the absorption efficacy per sheet of Brand B to Brand A is 1/1.5, Brand B will absorb $\dfrac{4}{3.5} \times \dfrac{2}{1} \times \dfrac{2}{1.5} \times \dfrac{1}{1.5} = \dfrac{128}{63}$ more water than Brand A. So Brand B is about twice as good for this task. This coincides with our previous result of the ratio of rolls needed 4.9/2.4. With this alternative approach of multiplying ratios it would be easy to make adjustments in case the length of the sheets or the price was not the same for both brands. All we would have to do is to multiply the previous product of ratios by the ratios of the prices, and by the ratio of the length of the sheets. In these cases, as with the thickness of the sheets, we would be dealing with inverse ratios.

To find how many rolls of Brand B we would actually need, we can find the number of sheets in a roll, using the average circumference (4π), multiplying it by the number of layers that fit in the usable width $(2 \div \dfrac{1}{64})$, and dividing by the length of the sheets (10). So the number of sheets is $4\pi \times 128 \div 10 \approx 161$. (Notice that this result is very close to the result obtained with the other method.) Because each sheet of Brand B absorbs one ounce of water, this is also the number of fluid ounces that each roll can absorb. The total number of rolls is $\dfrac{384}{161} \approx 2.4$.

Exercise 3. Derive formula (1) as the limit of polygonal rings formed by trapezoids (see Figure 4).

Exercise 4. Discuss in what ways is formula (1) analogous to the formula for the volume of a torus obtained by rotating a circle around an axis outside the circle. The volume of the torus is equal to the product of the area of the circle times the circumference traced by its center.

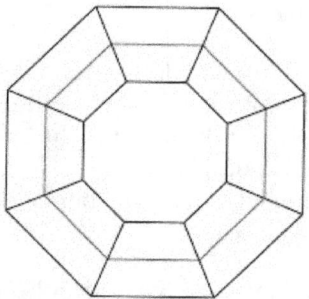

Figure 4. A ring formed by trapezoids.

Concluding remarks

When teachers pose a mathematical problem to their students they often do so because the problem can be solved with a mathematical approach that teachers want to illustrate. In the problem above, the intent of the teacher was that students had an opportunity to use an arithmetic series and the formula to find its sum. Problems can be excellent ways to foster the development and understanding of particular mathematical concepts and procedures. However, students may use an alternative solution process that does not require the concept or process the teacher wants to emphasize. Teachers need thus to be aware that alternative solutions may be found by their students that do not involve those concepts or procedures. In that case, the teachers needs to decide at what point, and to what extent to discuss the alternative approaches. It is important that teachers look at problems they pose from multiple perspectives, and try to foresee alternative solutions. That way teachers can better plan how and when to use alternatives so that it becomes an enriching experience for all the students, rather than becoming a situation where some students had the opportunity to develop their thinking with respect to specific mathematical concepts and method and others did not. Of course, sometimes students may surprise us and find an approach we did not foresee.

Learning to see the easy is one of the possible benefits of looking back at a problem and reflecting on its solution. Finding a simpler solution does not mean that our original approach is less valuable. The first method that occurs to us very likely gave us some insight into mathematical relations of a certain kind in the given situation, and perhaps uses mathematical ideas that are freshest in our mind. Furthermore, often we find a simpler path only after we were able to solve the problem in another way. By taking time to consider alternatives once they have found a solution, students may find an easier solution. Students may realize it is not always necessary to apply the most complicated math concepts that they know to solve even what appear to be difficult problems.

However, even when we find a simple solution first, it is worthwhile to take a second look at the problem and use a different proof. The second solution may give us a different kind of insight. As Polya points out, there are also other benefits of looking back, such as establishing connections. A few connections were hinted at above, but a full treatment would go beyond the main focus of this paper.

Of course, Polya and De Finetti are not the only authors to emphasize the importance of reflection when solving problems. Shulman states that "the more complex and higher-order the learning, the more it depends on reflection—looking back—and collaboration—working with others." (Shulman 2004, p. 319). The importance of reflection is not restricted to mathematics learning. Shulman also describes how studies of expertise in the solving of physics problems indicate that the most able problem solvers do not learn by just doing, that they do not learn from simply practicing the solving of physics problems. They learn from looking back on the problems they have solved and learn by reflecting on what they have done to solve them. Able problem solvers learn, not just by doing, but by thinking about what they do. (Shulman, 2004, p. 319).

Good teachers understand and convey to their students the benefits of looking back at a problem. Learning to see the easy is one of them.

References

Bramly, Serge (1994). *Leonardo: The artist and the man.* New York: Penguin Books.

De Finetti, Bruno. (1967). *Il "saper vedere" in matematica.* Torino, Loescher Editore.

Heydenreich, Ludwig Heinrich. (1954). Leonardo da Vinci. New York: Macmillan.

Polya, George. (1973). *How to solve it: A new aspect of mathematical method.* (2nd edition). Princeton, NJ: Princeton University Press.

Shulman, Lee S. (2004). Teaching alone, learning together: Needed agendas for the new reform. In Shulman, Lee S. *The Wisdom of Practice: Essays on Teaching, Learning, and Learning to Teach* (p. 309-333). San Francisco, CA: Jossey-Bass.

6 MULTICULTURAL AND GENDER EQUITY ISSUES IN A HISTORY OF MATHEMATICS COURSE[6]

In this article we discuss how to address multicultural and gender equity issues in a history of mathematics course designed for prospective secondary mathematics teachers. We also include an activity for high school students in which they can see that contributors to mathematics are not limited to European males. The first author has taught the course four times and the second author, who developed the activity, was a student in the course in the Spring semester of 2012.

What do mathematicians look like?

Some years ago one of the student teachers in our program devised and conducted a high school mathematics lesson that was different from most mathematics lessons. She showed the "human face" of mathematics. The lesson incorporated questions about the lives and some of the major contributions of several famous mathematicians. It was a very engaging lesson and the students participated actively. However, all the mathematicians in her lesson were male, European, and no longer alive. Some possible unintended consequences of such a selection came to mind. By including only male mathematicians would the teacher unwittingly send a message to her female students that mathematics is not for girls? By including only Europeans would students with different ancestries feel left out of mathematics? By including only dead people could we be sending the

---

[6] Flores, A. and Kimpton, K. E. (2012). Multicultural and gender equity issues in a history of mathematics course. *Journal of Mathematics Education at Teachers College*, 3(2), 37-42. Reprinted by permission.

message that mathematics is a thing in the past, rather than a lively and growing field of study?

It is not surprising that the student teacher chose male European mathematicians for her activity. Looking at mathematics textbooks one can verify that the vast majority of mathematicians mentioned are male and European. For example, when I (Flores) was a student at the university, my calculus book had a list of biographical dates (Courant and John, 1965). All 59 people listed were male and 56 of them were European. The exceptions were Euclid, Ptolemy and Michelson. Present day textbooks do not fare much better. For instance, in the index of a reform calculus textbook (Hughes-Hallett et al., 2009) there are 21 people listed with years of birth and death. All of them are male. Simmons (2007), in his book with 33 brief lives, includes only one biographical sketch about a woman (Hypatia). Except for the ancient mathematicians who lived in Alexandria (Euclid, Heron, Pappus) all the rest are European.

This focus on male European mathematicians may be one of the reasons why some students perceive mathematics as a male dominated field or that it is not for all segments of the population. The title of the book *"Multiplication is for White people"* (Delpit, 2012) reflects the belief that students from some ethnic groups have that mathematics is not for them. Future secondary mathematics teachers need to be aware of such perceptions and of the possible sources that may cause them, in order to provide their students of a more balanced picture of who can participate in and contribute to mathematics.

Female and non-European mathematicians in the history of mathematics

At the University of Delaware, future mathematics teachers take a history of mathematics course the year before student teaching. The main purpose of the course is to prepare them to become better mathematics teachers. This is done several ways. One is by developing their own knowledge of mathematics for teaching through the history of mathematics (Huntley and Flores, 2010). Another purpose of the course is to show prospective teachers that contributions to mathematics have come from individuals, both male and female, from many different cultures and ethnic groups.

In this course, each of the participants becomes the expert in one of the historical topics presented in the textbook. Students make two presentations to the class related to their topic and write three papers about the topic from different perspectives that are then combined into a final paper. One of the papers is a biographical sketch about a mathematician who made important contributions to their topic.

The first year I taught the course I encouraged students in the class to choose females and non-European mathematicians for their biographical sketches. However, even though the majority of the students in the class were female (15 females, 10 males), all the mathematicians chosen for the biographical sketches were males. Next year, I provided students with a list of 30 female mathematicians with a brief description of their contributions to several topics, as well as about 20 references with biographical information about women in mathematics. I highlighted the fact that there are a growing number of women who have made important contributions to mathematics as researchers, teachers, commentators, editors, or disseminators. I mentioned in class that the previous year everybody had chosen a male mathematician. Again, I encouraged them to choose females and non-Europeans. Even though there were 21 females and 5 males in the class, only one student chose a female mathematician (Sophie Germain) for her biographical sketch. In terms of cultural diversity the class did not do better either. Only one student chose a non-European mathematician.

I shared these results with a colleague at another university who is also interested in ways of improving the gender equity of topics presented in her classes. She suggested a somewhat drastic approach. Maybe one semester I could let students choose only from female or non-European mathematicians. She said that it could be argued this is fair because most mathematics books and mathematics classes only mention the European males, and one purpose of the course is to broaden their horizons (Jensen, 2010). She also suggested having students write about living mathematicians, and use that to emphasize the current diversity in the mathematical sciences. I thought about these suggestions, and whether to ask students to write two biographical sketches, where at least one of the biographies should be about a person who was either female, or non-European, or who was still alive during their own lifetime. I discussed the idea with one of my colleagues at Delaware. She thought that requiring two biographies would "allow students to do one of their own choosing and one within the 'confines' of the additional parameters..., giving them a bit more freedom" (Bartell, 2010). So next year I changed the assignment to writing two biographies.

In the last two years, due to the modified requirements for the biographical sketches there was more diversity represented in the people students wrote about. This year, for example, there were 6 males and 14 females in the course. Eighteen students chose a European male mathematician for one of their biographical sketches. Fifteen students chose non-European male mathematicians (among these, six from the United States). Seven students chose a female mathematician (five of them from Europe). In terms of recent contributors to mathematics, six students chose mathematicians who were alive during their own lifetime.

Cultural and gender bias in the textbook

Although it is encouraging that there is more diversity in the list of biographical sketches, there are still some issues that need to be addressed with future teachers as they learn to use teaching tools like textbooks. Future teachers need to realize in what ways textbooks, often unwittingly, may be contributing to perpetuate biased perceptions of mathematics by focusing almost exclusively on contributions of males, Europeans, and people who are no longer alive.

The textbook used in the history of mathematics course pays "most attention to the story of those parts of mathematics that we teach and learn in school" (Berlinghoff and Gouvêa, 2002, p. 5). The mathematics that we teach in the United States has its roots in contributions from Ancient Egypt and Mesopotamia, Greece, India and the Islamic empire. This tradition later continued and flourished in Europe. As a result, in the textbook, some traditions and cultures "receive less attention because they have had much less influence on the mathematics that we now teach" (p. 5). However, many of the topics that we teach in schools nowadays were independently discovered by people from around the world. These discoveries often pre-date the European mathematics that had a direct impact on our curriculum. For example, the Pythagorean theorem and symmetry patterns can be studied using art from Africa (Gerdes, 1999; Zaslavsky, 1973) or other parts of the world. Future teachers can also benefit from a more complex view of how people in different parts of the world interacted with each other and enriched and disseminated each other's mathematical knowledge to produce a wealth of knowledge that eventually made possible the rise and development of modern European mathematics (Joseph, 1992).

Early in the book, the authors of the mathematics history textbook address the issue that few women are mentioned in the book. The textbook indeed mentions many more males than females. Only 16 females are listed in the index of the book, in contrast to about 226 males. Berlinghoff and Gouvêa explain that before the 20th century in many places women were denied access to significant formal education in the sciences. In an early session of the course, in a discussion conducted by the two of us, we addressed the

78

role of women in mathematics in the past and at present. We invited students' input to the following questions:

> What is your position on the authors' statement that in the past mathematics achievement of women have been "obscured"? (p. 6) Do you agree with the statement that "in our times most of the barriers to women in the sciences have been dissolved" (p. 6). Why or why not? In what ways does your own personal experience coincide or contradict what the authors say?

Many of the girls in the class said that they were encouraged by their mathematics teachers in high school to continue their mathematics career. This was true for both their male and female teachers. However, gender differences in terms of expectations still exist. A boy with mathematical talent is likely to be encouraged by family and teachers to become an engineer rather than a teacher. All the students in the class, male or female, agreed that their high school textbooks left out many women mathematicians, and that was a major reason why they could not recite many names of female mathematicians when we asked. Furthermore, a survey of the research literature on girl's confidence in mathematics conducted by the second author shows that even today there are also other factors that affect girl's confidence in mathematics differently from boys, such as the competitive nature present in the mathematics classroom and curriculum (Niederle & Vesterlund, 2010), the onset of adolescent puberty (Buerk and Oaks 2001), and the availability of mentors and role models (Rogers & Kaiser 1995).

Berlinghoff and Gouvêa also mention that the "perception that mathematics is a male domain has been a remarkably resilient self-fulfilling prophecy" (p. 6). However, by listing so few women who have contributed to mathematics, textbooks may be helping to perpetuate this perception. There are several chapters in the book, such as statistics and computer science, where more women could have been easily mentioned. Future teachers need to be aware of that, and they can provide a more balanced picture. Although students in the class chose female mathematicians in only seven out of forty possible choices, this can still be viewed as an improvement compared to the gender ratio in the textbook (16 to 226).

Relating the history of mathematics to the high school curriculum

Another course assignment students were asked to do is to develop activities that relate their historical topic to the teaching of mathematics in today's high schools. I (Kelly) designed an activity incorporating information about contributions of different statisticians. Referencing the class textbook, "Math Through the Ages," I first presented in class a brief synopsis of real contributions to the field of statistics from various mathematicians. This slide presentation allowed me to illustrate the history of statistics visually. Students learned about the works of John Graunt, the first to see patterns in numerical burial records in 1662, Abraham de Moivre, who in 1733 discovered the importance of the normal distribution, Sir Francis Galton, who developed regression and correlation to help study human heredity and genetics, as well as other statisticians. I then gave my fellow students an activity in which they had to match the name of the statistician with his/her corresponding quotation. This activity, named "Celebrity Statisticians" (included at the end of the article), creates a fictional situation where statisticians from different ages are interviewed in the present day and make outrageous comments. Through this activity, I hoped to increase student motivation by putting historical statisticians in a more contemporary context. The lives of today's pop-singers, Hollywood actors, and professional athletes are heavily featured in today's media. Thus, most, if not all, adolescents have a famous celebrity that they look up to and idolize. So why not transform historical statisticians into modern-day celebrities? Each comment, although invented by me, is based on the actual work of the statistician. Creating clever quotations that fuse together modern day references (Justin Bieber, Madison Square Garden) with past mathematical contributions makes each statistician much more relatable in the classroom.

Looking back on my lesson, I found that this was a very effective activity because it presented the pioneers of statistics in a new light. It made students search for the underlying clues that reinforced what they learned in my initial presentation. For example, Gertrude Mary Cox should be paired with the comment, "I could have sold out Madison Square Garden faster than Justin Bieber, if I hosted my lecture conferences on economics, biological and nutritional problems, and plant and animal science there." With Justin Bieber's popularity, most adolescents can recite his life biography, let alone tell you how fast his concert sells out Madison Square Garden. Students should be able to piece together the parallelism between Bieber and Gertrude Cox. While Bieber holds pop-concerts, Cox held many popular mini-lectures and conferences that applied statistics to other life disciplines.

The list of nine names in the activity includes two women and one African American statistician. However, in the first version of the activity all nine people were European males. It was not until the instructor pointed this fact out that I realized the activity could have some unintended messages. One of my interests is gender equity in mathematics, and during the semester I did a literature review on the topic as an independent study, with special emphasis on strategies to encourage girls' participation in class. One of the four main factors I focused was the availability of mentors and role models. In the report of my study I pointed out that especially when teachers discuss the history of mathematics, many vital female mathematicians are left out of the picture. I found this part of my research very interesting because it relates to what I noticed in our History of Mathematical Ideas course. I found that in the textbook the history of mathematics is dominated by White, European males. If teachers neglect to incorporate female role models and mathematicians into their instructional lesson, it automatically makes femininity a powerless force in mathematics. When girls do not see relatable mathematicians within the pages of their textbook and their teacher does not discuss their exclusion, then the stereotype that 'math is a male's domain' is forever strengthened (Sadker & Sadker, 1995). A lack of role models inside the classroom creates a "culture of mathematics [that] remains distant, cold, and undesirable for too many women" (Rogers & Kaiser, 1995, p. 15). So it was especially interesting that my initial list of statisticians had only males. The reason is that my activity was based on the information provided by the chapter on the history of statistics in the textbook, in which all statisticians mentioned are males. From this activity, I realized how important it is that future teachers are not only aware of gender biases in textbooks, but that we create classroom activities with a conscious effort. Our students deserve a well-rounded mathematics education that provides a balanced picture of who can make contributions to mathematics. To promote girl's confidence inside the classroom, teachers need to rewrite women back into their instruction. By discussing how women have made significant strides in history we can show that mathematics is for everyone.

Final remarks

Exposing the history behind mathematics shows that it is never a field studied in isolation, but it has its own story and tradition. It sears open a new life that helps students see mathematics as an integrated subject. A history of mathematics course can effectively portray mathematics through a historical means. Students in the class researched the lives of its individual contributors and became aware of its underlying culture. In return, students learned how to employ more creative, yet engaging activities in the classroom. When students take on a more active status in the classroom, they gain a better foundation for the information presented. The way this course was conducted did also provide students with a wider academic scope that has expanded their teaching horizons. It has given new instructional strategies and ideas that can help future teacher create more cultural and gender equity in the classroom. This is an important component in the preparation of future teachers.

References

Bartell, T. (2010). Personal communication. May 7, 2010.

Berlinghoff, W. P., & Gouvêa, F. Q. (2002). *Math through the ages: A gentle history for teachers and others*. Farmington, ME: Oxton House Publishers.

Buerk, D. & Oaks, A. (2001). Empowering young women in mathematics through mentoring. In W. G. Secada, J. E. Becker, J. Rossi, G. F. Gilmer (Eds.), *Changing the faces of mathematics: Perspectives on gender* (p. 107-115). Reston, VA: National Council of Teachers of Mathematics.

Courant, R., & John, F. (1965). *Introduction to calculus and analysis* (Vol. 1). New York: Wiley.

Delpit, L. D. (2012). *"Multiplication is for White people": Raising expectations for other people's children*. New York: New Press.

Gerdes, P. (1999). *Geometry from Africa: Mathematical and educational explorations*. Washington, DC: Mathematical Association of America.

Hughes-Hallett, D., Gleason, A. M., McCallum, W. G., & et al. (2009). *Calculus: Single and multivariable* (5th ed.). New York: John Wiley.

Huntley, M. A. and Flores, A. (2010). A history of mathematics course to develop prospective secondary mathematics teachers' knowledge for teaching. *PRIMUS, 20*(7), 603-616.

Jensen, J. (2010). Personal communication. March 25, 2010.

Joseph, G. G. (1992). *The crest of the peacock: Non-European roots of mathematics*. London: Penguin Books.

Niederle, M. & Vesterlund, L. (2010). Explaining the gender gap in math test scores: The role of competition. Journal of Economic Perspectives, *24*, 129-144. Retrieved February 21, 2012 from http://www.stanford.edu/~niederle/NV.JEP.pdf

Rogers, P. & Kaiser, G. (1995). *Equity in mathematics education: Influences of feminism and culture*. London: RoutledgeFalmer.

Sadker, M. & Sadker, D. (1995). *Failing at fairness: How our schools cheat girls*. New York: TouchStone, 1995.

Simmons, G. F. (2007). *Calculus gems: Brief lives and memorable mathematics*. Washington, DC: Mathematical Association of America.

Zaslavsky, C. (1973). *Africa counts: Numbers and pattern in African culture*. Boston: Prindle, Weber & Schmidt.

Celebrity Statisticians
They said what??

Name: _____

Directions: Nine statisticians were caught saying these outrageous comments during their interviews on E! Match up the historical statistician to the quotation that he/she would most likely have said.

John Graunt

"Statistics is such a normal part of everyday life! How can anyone not use it?"

Edmund Halley

"The likelihood that research workers started using statistics because of me is high."

Gertrude Mary Cox

"I could have sold out Madison Square Garden faster than Justin Bieber if I hosted my lecture conferences on economics, biological and nutritional problems, and plant and animal science there."

Abraham de Moivre

"Haters can hate. Our physical and mental characteristics can be mapped to the normal curve!"

David Blackwell

"Some of these statisticians think they had it so hard. I had to also fight racial discrimination to become a credited statistician. My hard work made me become the first African American mathematician to be inducted in the National Academy of Science!"

Lambert Quetelet

"Men say that math is a male's domain. Ha! As a nurse, I saved so many of their lives during war using statistical charts, tables, and graphs!"

Sir Francis Galton

These guys are getting too much credit. After all, I was the first mathematician to analyze and understand my data!"

Florence Nightingale

"Insurance companies are forever in my debt. I am the father of actuarial science!"

Ronald Aylmer Fisher

"Statistics was the best fit for me. After all, I did invent regression and correlation!

83

7 EMPIRICAL APPROACHES TO THE BIRTHDAY PROBLEM[7]

Many students do not have systematic opportunities to develop probabilistic intuitions. Developing a sense of the distribution of random outcomes requires at least two things. First, students need a large number of experiences with the same probabilistic situation, which is a rare event in most classrooms. Second, they need to encounter a variety of probabilistic situations including equally-likely events (such as tossing a coin or one die) and unequally-likely events (such as rolling various sums with a pair of dice). Often students form their conceptions about probability based on a very limited number of experiences, and they are frequently not aware that they have misconceptions and poor intuitions about probabilistic situations. It is important to bring these poorly formed notions to the conscious level so that students can modify them. One way to do this is to have students make and justify predictions about a probabilistic situation where the results will be surprising and unexpected. This allows any misconceptions to become explicit. When students articulate their ideas before they experiment or analyze the situation, they maximize the potential that such situations will make them rethink their basic assumptions (National Council of Teachers of Mathematics [NCTM] 1989, p. 110).

One approach to develop understanding of basic concepts of probability is to use simulations to construct empirical probability distributions (NCTM 2000, p. 324). The use of technology can facilitate students' learning of

---

[7] Flores, A. and Cauto, K. M. (2012). Empirical approaches to the birthday problem. *Mathematics Teacher*, *106*(2), 134-137. Copyright National Council of Teachers of Mathematics. Used by permission.

probability in at least two ways. First, students can generate a large number of simulations in a short time, so that they can observe the variability from one experiment to the next. Second, because the samples are generated fairly easily by the computer or the calculator, students can focus their attention on analyzing the data (NCTM 2000, p. 254). Students can thus perform experiments to develop probabilistic intuitions and concepts before dealing with theoretical probabilities. Analyzing the theoretical probability in light of the empirical results allows students to see the relationship between the two.

An unexpected situation: At least one repeated birthday in a group

Students find it surprising when in a given classroom two students have the same birthday. When students are asked what is the probability that in a group of 40 persons at least two have the same birthday (same month and day, not necessarily same year), many of them think the probability is fairly low given that 40 is a small number compared with 365. Although students see fairly easily that in order to be certain that two people have the same birthday you need 366 persons in the group (year with 365 days), they have not developed an appropriate intuition about the likelihood of having at least one repeated birthday in a certain size group.

In this article we will describe two activities in which students conduct experiments with random numbers so they can see that having at least one repeated birthday in a group is not really that unusual. The first empirical approach was conducted by Kevin Cauto in a secondary math methods course, and his fellow students participated as if they were secondary school students. The second empirical approach was used by Alfinio Flores with in-service teachers. The teachers participated in the activities pretending they were secondary students. In a third activity students use a calculator program to deal with the theoretical probability.

We will make several assumptions to simplify the experiment. We will use the year with 365 days, thus disregarding birthdays on February 29. We will assume that each day is equally likely for birthdays. In real life this is not quite true; in the U.S. the daily average of births is slightly higher during the summer months July - September, and lower in January (James, 2005). Rather than listing the birthdays by month and day, we will deal with numbers between 1 and 365. Thus 2 corresponds to January 2nd, 32 corresponds to February 1st, and 365 corresponds to December 31st.

First empirical approach

Before students do the activity, let students guess the probability that in a group of 23 people there is at least one repeated birthday. Write their guesses on the board. To estimate the probability each student in the class will do an experiment, and based on the results of the whole class they can compute the empirical probability.

Students can use a computer or a graphing calculator to generate lists of random numbers. For example, with the calculator TI-84 Plus, pressing the **MATH** key, and moving the cursor to the **PRB** (probability) menu, the option **randInt(** will be shown. For this function we can choose the range of random numbers and how many will be generated. So for example, **randInt(1, 365, 23)** will generate 23 random whole numbers between 1 and 365 (inclusively). We can instruct the calculator to store the numbers in a given list by using the **STO>** key. So by typing

randInt(1, 365, 23) STO> L2   the calculator will generate 23 random numbers and will store them in list 2. Once the data are stored in a given list students can order them. To do so, students can press the **STAT** key and in the **EDIT** menu choose **SortA(** and enter the list they want to sort, for example, **SortA(L2)**. In Figure 1 two lists have been generated. After sorting, we see that list L1 has one entry repeated, 97. That means two people had the same birthday (97 corresponds to April 7).

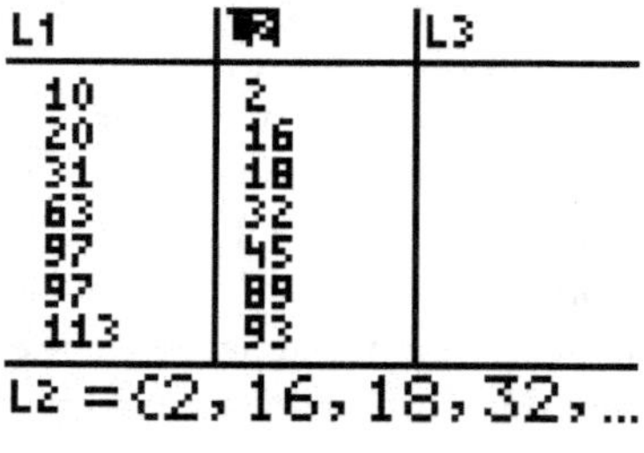

Figure 1. Two lists of ordered birthdays

First simulation: 10, 20, 31, 63, **97, 97**, 113, 122, 136, 152, 169, 179, 192, 212, 213, 222, 228, 294, 332, 342, 354, 360, 363 (repeated birthdays highlighted)

Second simulation: 2, 16, 18, 32, 45, 89, 93, 94, 99, 103, 111, 112, 117, 150, 220, 245, 283, 290, 299, 309, 310, 320, 350 (no repeated birthdays)

Students in a classroom can generate, store, and sort their own lists. As students go through their lists, they can see whether there are repeated numbers or not. When the second author conducted this activity with a group of twenty students, each of them generated a list of 23 numbers, representing the birthdays in a group. It turned out that in 11 of the 20 cases there was at least one repeated birthday in the group. The participants were quite surprised that in so many simulations (55%) there was at least one repeated birthday.

Increase the size of the groups, say, to 40, and let students guess again the probability that in a group of 40 there is at least one repeated birthday. Write their guesses on the board. To simulate the situation with the calculator students would need to modify slightly the instruction to **randInt(1, 365, 40) STO> L2**, then sort the list and verify whether there are any repeated numbers. The results for the whole class will serve again to estimate the empirical probability.

When students generate longer lists it becomes obvious that the probability of at least one repeated birthday for the given group size is fairly high. For example, when 20 students generate lists of birthdays for groups of size 40, they find that in the vast majority of the cases there is at least one repeated birthday. In one case, twenty simulations of groups of size 40 yielded 18 cases of at least one repeated birthday, and only two groups with no repeated birthdays. The empirical probability is thus 90%.

Before doing the experiments very few students guess that for groups of only 23 students the probability is already about 50% that two individuals will have the same birthday, or that the probability of at least one repeated birthday in a group of 40 is almost 90%. For them, 23 is too small compared with 365 to think that the probability is about 50%. Because of the surprising nature of the results, it is convenient to solidify students' understanding of the probabilistic situation through another simulation.

Another empirical approach

In the next activity, students run a program that simulates adding people at random to a group until there is a repeated birthday. For a given group the program randomly adds a new birthday to the group and records it until there is a repeated birthday. The calculator will display how many people were in the group when a birthday was first repeated, and does this for 50 groups. The calculator commands are on the left; the explanation of each line on the right.

PROGRAM:CUMPLE1

For(N,1,50,1)	start of loop for 50 groups
365→dim(L4)	list for 365 birthdays
Fill(0,L4)	zero for each day
0→P	zero people in the group
Repeat max(L4)>1	loop group; instructions executed until one

	birthday is repeated
$(1+\text{int}(365*\text{rand}))\rightarrow R$	new birthday added at random
$(1+L4\ (R))\rightarrow L4\ (R)$	one is added to birthday tally
$1+P\rightarrow P$	number of people is counted
End	end of loop for group
Disp P	number of people in group displayed
Pause	pause between groups (press ENTER to continue)
End	end of loop for 50 groups

The table below shows the results of running the program. We see that in this sequence of 50 experiments in all cases it took 45 or fewer people to have a repeated birthday. Of course, running the program again will give slightly different results. Students can determine the median of the data in the table (22). This is another way to see that the probability of repeated birthdays for groups of more than 23 is bigger than 50%.

Table 1. Frequencies for groups of different sizes

Number of people in group when birthday was repeated	Frequency
3	1
4	2
5	1
7	1
8	2
9	2
11	2
12	1
13	2
14	2
16	1
18	1
19	3
21	2
22	4
23	3
24	1
25	1
27	1
28	3
30	2
31	1

32	1
34	1
35	3
36	1
39	1
42	1
43	2
45	1

The graph below (Figure 2) represents the results of doing the simulation 200 times. Students can see that indeed in 107 of the cases (more than 50%) the groups were 24 or less when the repeated birthday occurred.

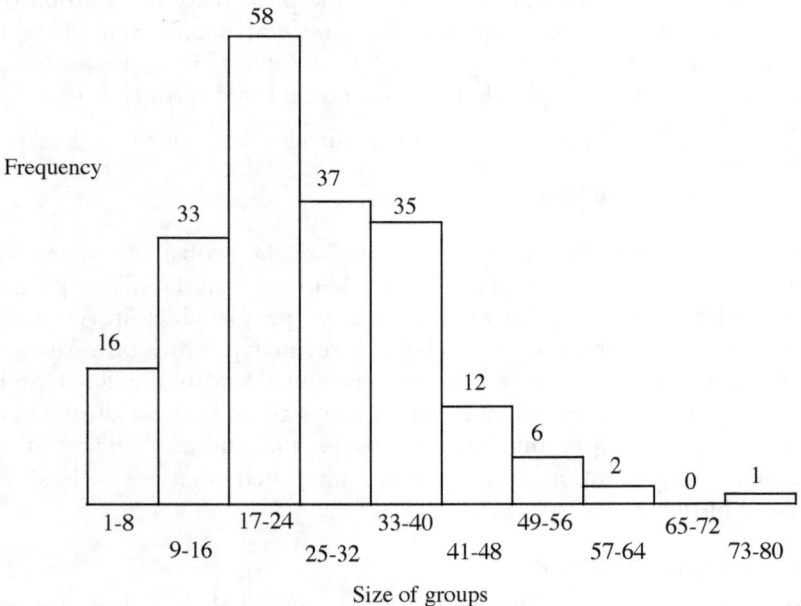

Figure 2. Distribution of size of groups when birthday was repeated

Theoretical probability

After students have done several simulations to see the experimental probability for having at least one repeated birthday in a group, they can deal with the theoretical probability of the situation. To find the probability of at least one repeated birthday in a group, it is easier to think first about the probability that there are no repeated birthdays among a group of people. That is, students will compute the probability of the complementary event first. If there is only one person, there will be no repeated birthdays, so the probability is 1. If there are two people, the probability of not having a repeated birthday is 364/365 because for this to happen, the birthday of the second person can be any of the remaining 364 days other than the birthday of the first person. If a third person is added, we need to multiply $\frac{364}{365} \times \frac{363}{365}$, that is, multiply the probability that the first two people have different birthdays times the probability that the birthday of the third person is one of the other 363 days different from the two birthdays. If a fourth person is added, we need to multiply the previous result, which is the probability that three people do not have the same birthday by 362/365, which is the probability that the new birthday is different from those already in the group. Thus, we can compute the probability of different birthdays in a recursive way. If we know the probability for a group of n people, we can compute the probability for $n + 1$ people multiplying the product corresponding to n people by $(365\text{-}n)/365$. If Q is the probability that in a group of $n + 1$ people there are no repeated birthdays, then Q = $\frac{364}{365} \times \frac{363}{365} \times ... \frac{365 - n}{365}$. Therefore, the probability of having at least two people with the same birthday is $1 - Q$.

Students can run a short program to calculate the probability of repeated birthdays for successive numbers of people in a group. Here is a program for the TI-84 calculator. For each one of 100 people added in succession, the program computes the probability that the new person's birthday is not the same as any of the people already accounted and multiplies it by the probability that there are no repeated birthdays among people already in the group. It displays the number of people and the probability of the complementary event in which we are interested, that is, at least one repeated birthday.

PROGRAM:BIRTHDAY

1→Q	Probability of no repeated birthdays for one person
FOR (N, 1, 100, 1)	Beginning of loop
Q*(365-N)/365→Q	Probability of no repeated birthdays adding one more person
DISP N+1, 1-Q, " "	Number of people, and probability of repeated birthday
Pause	Program pauses (hit ENTER to continue)
End	End of loop

Students can run the program and see what are the probabilities for successive numbers of people. In Table 2 we display the results for a few numbers, including 2, 4, 23, 30, 41, 50, 57, and 70. Students will see that for 70 people the probability of repeated birthdays is more than 99.9% even though 70 may seem relatively small compared to 365.

Table 2. Theoretical probabilities for repeated birthdays

Number of people in group	Probability of repeated birthdays
2	.002739726
4	.0163559125
23	.5072972343
30	.7063162427
41	.9031516115
50	.9703735796
57	.9901224593
70	.999159576

Concluding remarks

A common misconception about random samples is that the outcomes are "spread out" more or less evenly among the possible results. Thinking about 23 people compared to 365 days, students may think that there are a lot of slots available, and because on the average birthdays would be spread out two per month, that it is very unlikely that two people will have the same birthday. However, one of the understandings that students need to develop about random samples is that sometimes results within a given sample often are clustered or repeated, and that the distribution within a particular sample is not always uniformly or symmetrically distributed among the possible results. Doing simulations like the ones described above can help students develop such understandings.

References

James, Michael S. "Tables: Births and Deaths by Month, 1995 - 2002." http://abcnews.go.com/Health/Science/story?id=990641.

National Council of Teachers of Mathematics. *Curriculum and Evaluation Standards for School Mathematics.* Reston, Va.: National Council of Teachers of Mathematics, 1989.

National Council of Teachers of Mathematics. *Principles and Standards for School Mathematics.* Reston, Va.: National Council of Teachers of Mathematics, 2000.

8 ALGEBRA FROM CHIPS & CHOPSTICKS[8]

In these activities students represent triangular, square, pentagonal and hexagonal numbers with chips (see Figure 1), and they use chopsticks to break polygonal numbers into components. Students represent each part with an algebraic expression, then find an algebraic representation for the total, and thus establish relations between different algebraic expressions. This article focuses on the approaches students used to break pentagonal and hexagonal numbers into lower polygonal numbers. Of course, other decompositions of polygonal numbers are possible and are hinted to at the end of the article. The goals of these activities are to help students

- find number patterns that have geometrical structure;
- develop their own strategies to count numbers that form a pattern;
- compare their strategies and learn from each other; and
- experience the beauty of mathematic

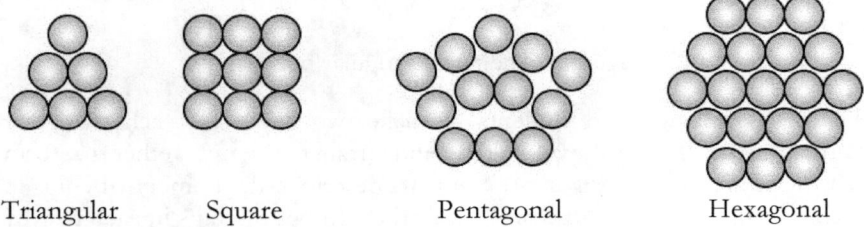

Triangular   Square   Pentagonal   Hexagonal

Figure 1. Polygonal numbers

---

[8] Yun, J. O. and Flores, A. (2012). Algebra from chips & chopsticks, *Teaching Mathematics in the Middle School*, 17(6), 324-331. Copyright National Council of Teachers of Mathematics. Used by permission.

The activities were developed by the first author (Yun 2007) and conducted with six groups in a public magnet secondary school in Korea. The students are among the top 20% of their grade and have special talent for English and preference for social studies, but they are not especially gifted in mathematics.

For the first activity with triangular numbers the teacher asked students to represent the first few triangular numbers (Figure 2). Then she asked students to generalize to the n-th triangular number after investigating 1st, 2nd, and 3rd triangular numbers. In some classes the teacher gave students hints such as where is 2 in the second triangular and where is 3 in the third triangular number.

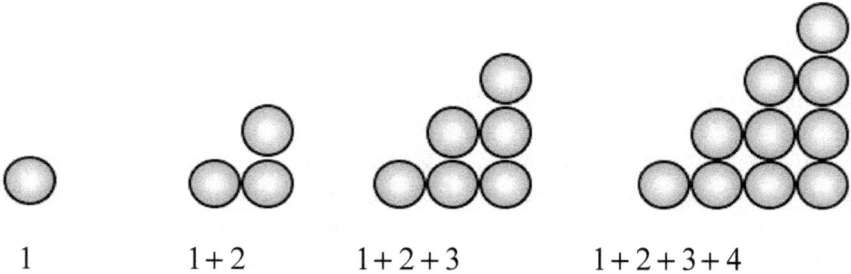

Figure 2. Triangular numbers as sums of consecutive whole numbers

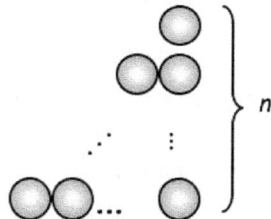

Figure 3. The n-th triangular number

The teacher then asked students to make two copies of each triangular number with chips of two colors, and arrange them together to form rectangular arrays (see Figure 4). Students described the number of chips in each rectangular array as the product of the number of chips in each row by the number of rows, and expressed the number of chips for the 2nd, 3rd, and 4th triangular number as half the chips in the corresponding rectangular array. They generalized to represent the n-th triangular number algebraically.

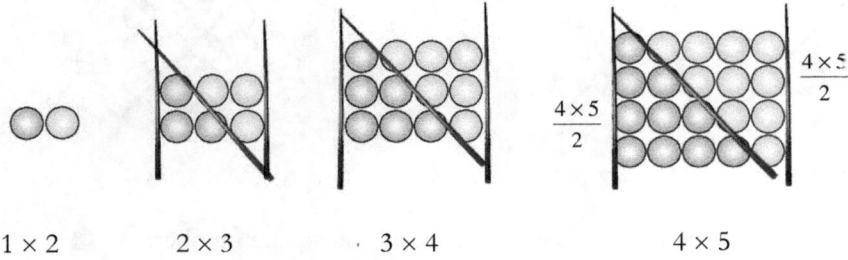

Figure 4. Rectangular arrays

$T_1 = 1 = (1 \times 2)/2$

$T_2 = 1 + 2 = (2 \times 3)/2$

$T_3 = 1 + 2 + 3 = (3 \times 4)/2$

$T_4 = 1 + 2 + 3 + 4 = (4 \times 5)/2$

$T_n = 1 + 2 + 3 + \dots + n = n(n+1)/2$

Then the teacher conducted an activity to represent square numbers as the sum of two triangular numbers using chopsticks. The arrays in Figure 5 are called square numbers. The number of chips in each row is equal to the number of rows. So, if there are n rows, the total amount of chips will be $n \times n = n^2$. The teacher asked students to use chips to build the first four square numbers. The teacher next asked students to find triangular numbers within the square numbers and use a chopstick to separate them (Figure 6).

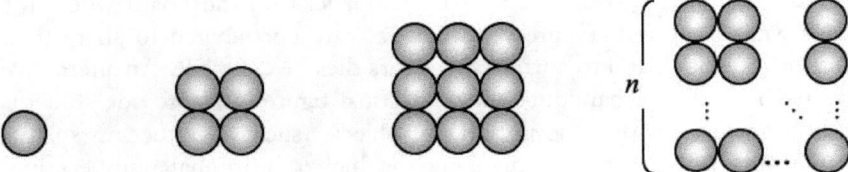

Figure 5. Square numbers

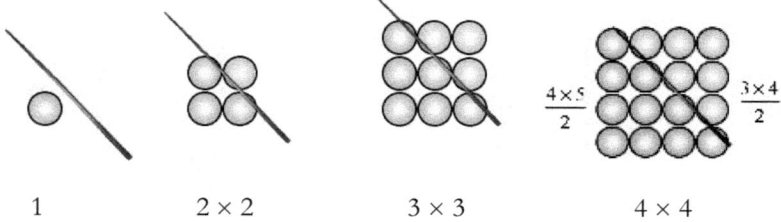

1   2 × 2   3 × 3   4 × 4

Figure 6. Triangular numbers and square numbers

The teacher asked students to express each square number algebraically as the sum of two terms that represent triangular numbers. For example,

$$S_3 = 3 \times 3 = (2 \times 3)/2 + (3 \times 4)/2,$$

$$S_4 = 4 \times 4 = (3 \times 4)/2 + (4 \times 5)/2.$$

Students generalized to the n-th square number, and verified that the algebraic expressions on both sides of the following identity are indeed equivalent. They did so by expanding and simplifying the right side.

$$n \times n = \frac{(n-1) \times n}{2} + \frac{n \times (n+1)}{2}$$

Then, after the teacher explained how to form pentagonal numbers with chips, she asked her students to find their own strategies for using chopsticks to partition the pentagonal numbers and finding numerical and algebraic representations of them. Then she asked students to do the same for central hexagonal numbers. The teacher prepared a slide presentation and asked students to draw imaginary chopsticks on the board where the slides were projected (Figure 7). Students were encouraged to share their ideas at the board, and to correct any errors their peers made. An alternative approach is to have handouts with polygonal figures and provide students with chopsticks or other long thin flat objects (such as fettuccini) so that participants can experiment breaking the higher polygonal numbers into lower polygonal numbers. This approach was used by the second author with K-12 in-service teachers in the U.S. Here we will describe different strategies used by the secondary school students.

Students first wrote the numeric representation of 2nd, 3rd, and in some cases 4th polygonal numbers, then wrote an algebraic representation. Students enjoyed finding many different strategies for just one question and even became competitive. In one class students found eight different ways to represent pentagonal numbers. In another class students were eager to share their strategies even after the bell had rung. Some of the students who did not show much interest at the beginning, after watching their peers present their novel ideas, became interested and tried to come up with their own ideas. In some cases students were amazed by the different strategies of their peers. Using tools to partition the arrays was useful in finding patterns. Although some students were not particularly fond of the chopsticks and wanted to use tools that could be bent like wire, thinking about where to put the chopsticks seemed to help students to concentrate on how to partition the array of chips.

Figure 7. Student expressing a pentagonal number as $(n-1)\times n+\dfrac{n\times(n+1)}{2}$

Pentagonal numbers

The arrays in figure 8 represent pentagonal numbers. The dotted lines have been added to emphasize how one pentagonal number is contained in the next. Each new pentagonal number is formed by adding a new layer consisting of three sides at the bottom, therby extending the two sides that meet at the upper vertex. This upper vertex has thus a different role than the other vertices in this kind of pentagonal array.

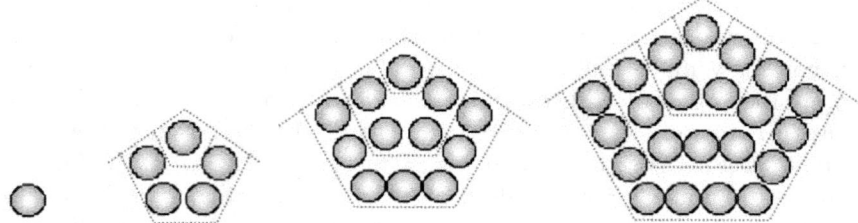

Figure 8. Pentagonal number

Pentagonal numbers can be decomposed into triangular numbers. Figure 9 shows one kind of decomposition. Students expressed each of the pentagonal number as a sum of three triangular numbers. For example,

$P_2 = 1 + 1 + 3 = 2 \times (1 \times 2)/2 + (2 \times 3)/2$

$P_3 = 3 + 3 + 6 = 2 \times (2 \times 3)/2 + (3 \times 4)/2$

Then they generalized the pattern to obtain a formula for the number of chips in the n-th pentagonal number.

$$P_n = 2T_{n-1} + T_n = 2 \times \frac{(n-1) \times n}{2} + \frac{n \times (n+1)}{2} = \frac{n(3n-1)}{2}$$

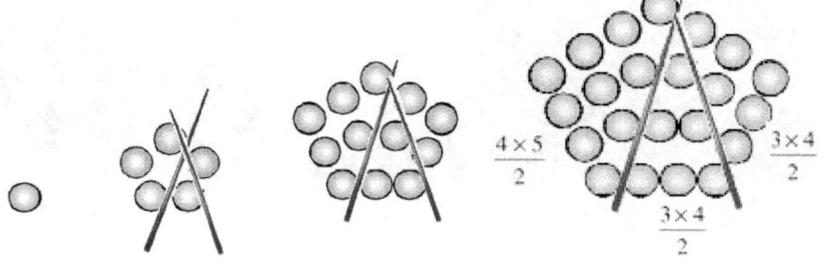

Figure 9. Pentagonal numbers and triangular numbers.

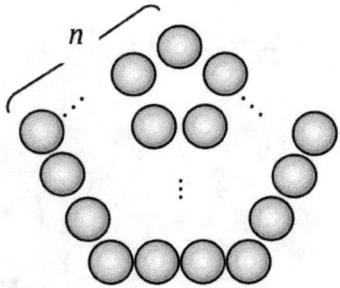

Figure 10. The *n*-th pentagonal number

Students can also break a polygonal number into a triangular number and a square number (Figure 11). In Figure 12, the pentagonal numbers have been squeezed to show more clearly how they are formed by a triangular number and a square number (Meavilla Seguí 2005).

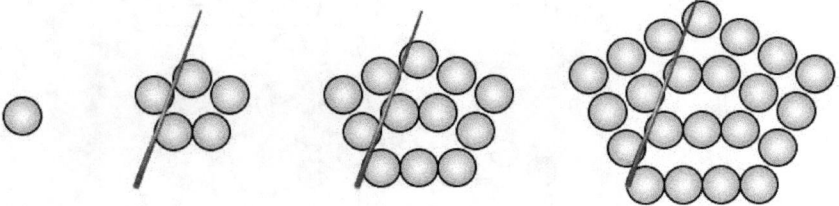

Figure 11. Pentagonal numbers as triangular numbers and square numbers

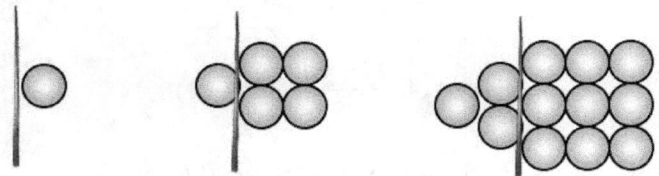

Figure 12. Pentagonal numbers squeezed

Students expressed the *n*-th pentagonal number as the sum of the (*n*-1)-th triangular number and the *n*-th square number. They verified that this algebraic expression is equivalent to the one obtained above.

Students found several partitions and the corresponding algebraic expressions (Figures 13 and 14). They verified that indeed the total number of chips is the same. For the partition generated by the dotted lines in Figure 8 students found this expression

$$1+\left(3\times 2-2\right)+\left(3\times 3-2\right)+\left(3\times 4-2\right)+\cdots=\sum_{k=1}^{n}(3k-2)$$

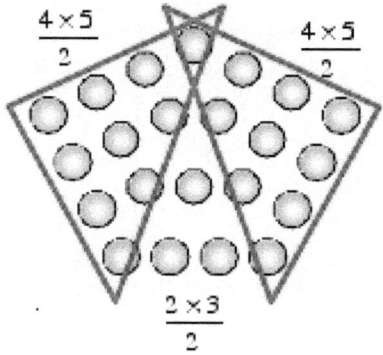

Figure 13. $\left(\dfrac{n(n+1)}{2}\times 2-1\right)+\dfrac{(n-2)(n-1)}{2}$

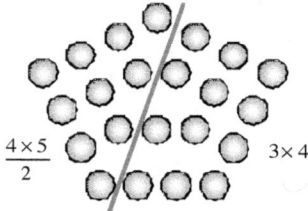

Figure 14. $\dfrac{n(n+1)}{2}+(n-1)n$

Students used "wire" for the following partition (Figure 15).

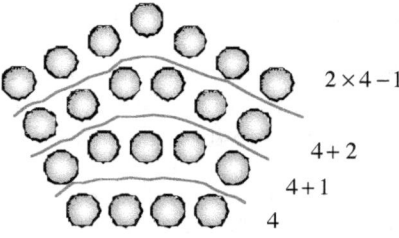

Figure 15. $n+(n+1)+(n+2)+...+(2n-1)$

Using the partition illustrated in Figure 15 a student realized that pentagonal numbers could be squeezed into trapezoidal shapes (Figure 16)

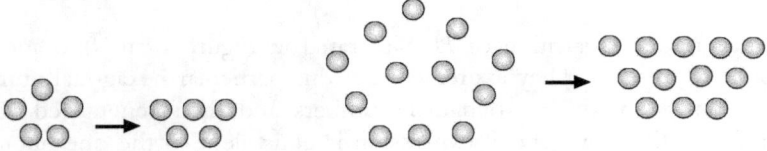

Figure 16. Another representation of pentagonal numbers

Hexagonal numbers

The arrays in Figure 17 are called the central hexagonal numbers

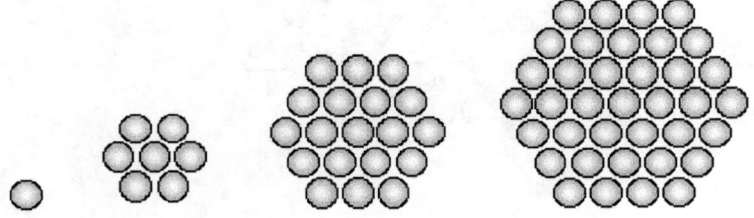

Figure 17. Hexagonal numbers

Students used chopsticks to break each hexagonal number into triangular numbers. They found several solutions. Figure 18 shows one.

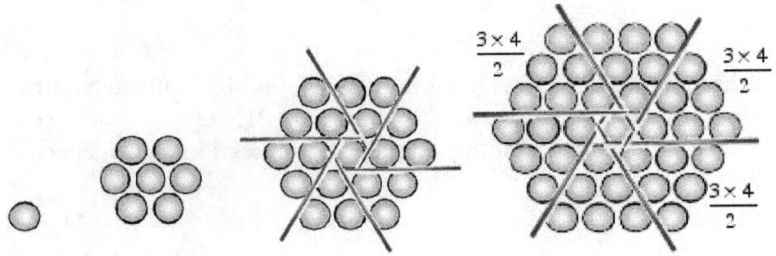

Figure 18. Hexagonal numbers and triangular numbers

Students expressed the hexagonal numbers in terms of the triangular numbers.

$H_3 = 1 + 6 \times (2 \times 3)/2$

$H_4 = 1 + 6 \times (3 \times 4)/2.$

After they wrote an expression for the number of chips in the fourth central hexagonal number, they generalized to the n-th central hexagonal numbers, $H_n = 1 + 6 \times n(n\text{-}1)/2$

Students used the partition of H_4 illustrated in Figure 19 to find another algebraic expression. They expressed first this particular hexagonal number as the sum of particular triangular numbers and then generalized. They verified that the new general expression is equivalent to the one obtained above.

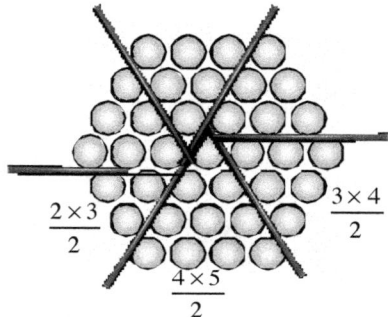

Figure 19. Another way to partition a hexagonal number

$H_4 = (4 \times 5)/2 + 4 (3 \times 4)/2 + (2 \times 3)/2$

$$H_n = \frac{n \times (n+1)}{2} + 4 \times \frac{(n-1) \times n}{2} + \frac{(n-2) \times (n-1)}{2}$$

Here are other partitions of hexagonal numbers found by students and the corresponding algebraic expressions (Figures 20, 21, 22). In each case students wrote the corresponding numerical expression with specific numbers first.

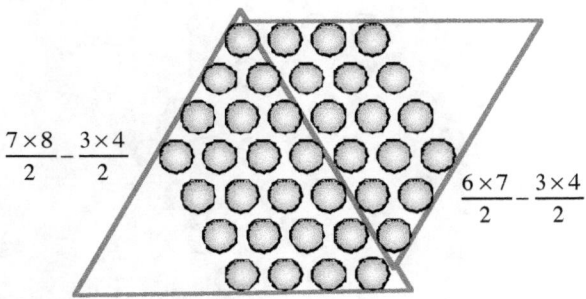

Figure 20. $\dfrac{(n+3)(n+4)}{2} - \dfrac{(n-1)n}{2} + \dfrac{(n+2)(n+3)}{2} - \dfrac{(n-1)n}{2}$

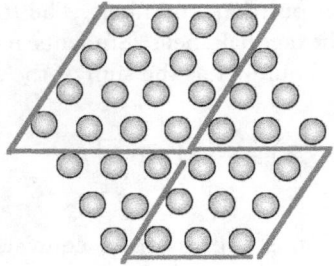

Figure 21. $n^2 + (n-1)^2 + \dfrac{(n-1)n}{2} \times 2$

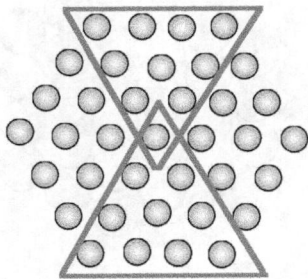

Figure 22. $\dfrac{n(n+1)}{2} \times 2 - 1 + (n-1)^2 \times 2$

A student found a beautiful decomposition of a hexagonal number as the sum of a triangular, a square, and a pentagonal number (Figure 23).

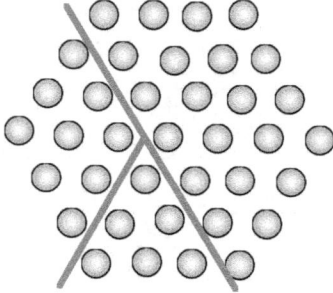

Figure 23. Triangular plus square plus pentagonal numbers

Figure 24 illustrates that polygonal numbers can be broken into other shapes that are not lower polygonal numbers. The fourth central hexagonal number is the sum of hexagonal shells. Students found an expression for the n-th central hexagonal number as the sum of the shells

$$1 + (6 \times 1) + (6 \times 2) + (6 \times 3) + \cdots = 1 + 6 \sum_{k=1}^{n-1} k \, .$$

Students can verify that this expression is equivalent to the expressions obtained above.

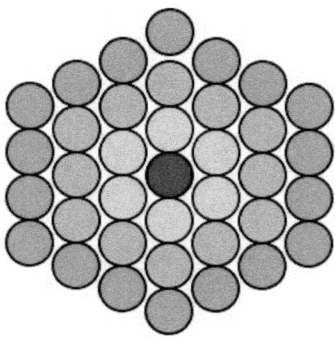

Figure 24. The fourth central hexagonal number

Final comments

Polygonal numbers are one kind of geometrical representation of numbers and relations among numbers. Students can use such geometrical representations as a means to explore algebraic ideas. With the help of these representations students can think about the relations among the numbers, express them using their own words, and represent them with letters. These activities can stimulate students to try to find various ways of solving a problem and appreciate the joy of finding various solutions. The activities also foster them to think how to find patterns, to express the patterns in numerical forms, and to generalize them into algebraic forms. A teacher can use geometrical representations to help students as they learn to use algebra to generalize and justify (Flores 2002). As we saw above, students used explicit expressions that stress the relationship among the parts such as $(1 \times 2)/2$, $(2 \times 3)/2$, $(3 \times 4)/2$, $(4 \times 5)/2$ for successive triangular numbers, rather than just writing the totals 1, 3, 6, 10. Students also referred these numerical expressions to corresponding geometrical representations within the pentagonal or hexagonal numbers. Using this approach as a scaffold, students were able to generalize to algebraic expressions using variables. Polygonal numbers and other geometrical representations used this way can thus provide a more concrete step towards the more abstract use of letters as variables or generalized numbers, which for beginners may be a little complicated.

References

Flores Peñafiel, Alfinio. "Geometric Representations in the Transition from Arithmetic to Algebra." In *Representations and Mathematics Visualization* edited by Fernando Hitt, 9-29. México: Departamento de Matemática Educativa del CINVESTAV-IPN, 2002.

Meavilla Seguí, Vicente. *La Historia de las Matemáticas como Recurso Didáctico*. Badajoz, Spain: Federación de Sociedades de Profesores de Matemáticas, 2005.

Yun, Jeong Oak. Polygonal numbers. Unpublished manuscript. Arizona State University, 2007.

9 THE "BUBBLE BOARD" AND CURVE FITTING[9]

The "Bubble Board" is a simple apparatus that allows students to simultaneously form 56 almost identical soap bubbles. The device was proposed by Rämme in 2001 as a tool for illustrating data collection and curve fitting. Here we describe a modified version we term the Bubble Board and explore the number of remaining bubbles over time and different mathematical models to fit the data. The board is made with a perforated slate of polycarbonate and clear drinking straws. One end of the straws is dipped in soap solution, and then the other end is submerged in a water tank, and the bubbles are formed (see Figure 1). As time goes by, some of the bubbles will burst and others will remain. Students can study the number of bubbles remaining at a given time. To facilitate the counting, soap bubbles can be made to last longer by adding small concentrations of glycerin to the bubble solution. Students can then study the relationship between time and the number of remaining bubbles for different concentrations of glycerin. With bubble board activities, students encounter various mathematical ideas, such as percentages and ratios, volume of spheres and cylinders, scatter plots, curve fitting using different models, exponential functions, and probabilistic models. In this article, we describe how the bubble board is made, as well as how bubbles are formed and the data collected. We also provide activities for students allowing them to fit curves to sets of data using linear, exponential, and logistic models. A complete list of materials and resources needed and places where these can be acquired are given in the appendices.

---

[9] Hammons, A. N., Flores, A., Pelesko, J. A., and Biehl, L. C. (2012). The "Bubble Board" and curve fitting. *Ohio Journal of School Mathematics*, No. 66, 9-16. Reprinted by permission.

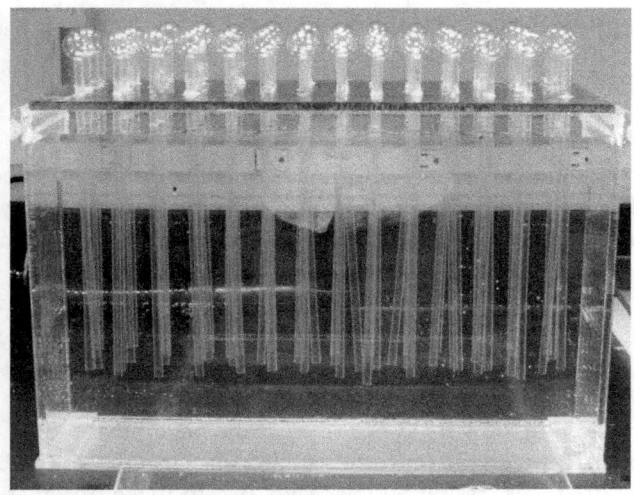

Figure 1. The bubble board.

Constructing the bubble board

The board is made of a rectangular piece of polycarbonate (Lexan) 0.5 cm thick, 16 cm by 38 cm, and has 4 rows of 14 circular holes, at a distance of 2.5 cm apart from each other. The diameters of the holes are 5.9 mm so that they can hold a straw with the same diameter tightly. If the fit is not quite tight, straws can be glued to the board. Teachers can drill the holes themselves, or they can download the design from the MEC Lab website and have the boards made (see Ponoko in Appendix B). It is best to use non-bending clear straws. When the board is ready, the straws, approximately 23 cm long and 0.59 cm in diameter, are inserted into the holes so that segments of 3 cm are on the side that will be dipped in the soap mix, and about 19.5 cm of each is on the side that will be submerged in water. The air displaced from the straws by the incoming water forms the bubbles. The straws will not be exactly the same length, so the part that is below may vary by one or two mm and some bubbles will be slightly larger in volume (about 1%) than others. The difference in radius of the spheres is hardly noticeable (about 0.3%). For purposes of dipping the straws in the soap solution, it is better to have all straw segments on that side the same length. A water tank about 26 cm deep, 38 cm long, and 11 cm wide is convenient to use with the board. An 8½ by 11 inch (21.6 by 28 cm) box frame or another shallow tray big enough for dipping the tips of the straws will be used. See complete list of materials in appendix A.

Making the bubbles

Students make three mixes (400 g each) of water and dish soap, 1% soap by weight, with three different concentrations of glycerin (2%, 4%, 6%). For instructions to make the mixes see Hammons 2009b. Then students pour one mix on a flat pan. For more uniformity in the duration of the bubbles across trials, have students wet the board before the first trial by inserting the board into the tank. Shake excess water from board over a sink and pat ends of straws with a dry paper towel. The shorter side of the straws is dipped into the soap mix. Then the board is pulled up so that each straw is coated with a soap film. The board is inverted and placed gently on top of a water tank so that the longer side of the straws is submerged. The displaced air will inflate the bubbles. Sometimes not all 56 bubbles will form, so students should note how many are there at the beginning. Students need to place the board very gently onto the tank to minimize having bubbles slide off the tops of straws. If any bubbles slide off, to make observation easier, students can simply pop those bubbles and then note the new staring number.

Relationship between percentage of glycerin and survival time of bubbles

Different factors affect the durability of soap bubbles, such as humidity of the air, temperature, volume of the bubble, temperature of water, water vapor inside the bubble, etc. (Behroozi and Olson 1994). In this activity, students keep most factors constant between experiments and vary only the concentration of glycerin. The length and diameter of straws is kept constant, so the amount of air displaced and volume of the bubbles is also constant. Other factors, such as humidity and temperature, that affect the durability of the bubble are assumed to be constant within the room during the experiment, although they can vary from one room to another and from one day to another. The concentration of dish soap is kept constant in the mixes. Students can repeat the experiments for different concentrations of glycerin, 2%, 4%, and 6%.

Collecting data

There are several ways in which students can register the time when a bubble bursts. Students can use on-line split timers (see appendix B for one) and choose the "Stop Watch" option. Once the bubble board has been inverted over the water tank, students click start, and press the "Split" button each time a bubble bursts. Data collection may be challenging as occasionally multiple bubbles pop simultaneously. Students should try their best to be accurate, but know that human error is part of any experiment and that reasonable results will likely be obtained despite this obstacle. To increase accuracy, have more than one student record data for each trial, with their eyes on the board and hand on the mouse to click a split as each bubble dies. When all bubbles have popped, or after the predetermined time, stop the timer. Students copy and paste the splits into Notepad or Word for later analysis. For each concentration, depending on the time available, students repeat the experiment at least two additional times (for a total of three) and up to nine total times. Students record how many bubbles were still standing after specified intervals of time into the Excel template. For solution A, students may find that the bubbles are popping too quickly to accurately record individual deaths. In this case, they can simply count the number of bubbles remaining at one minute intervals.

Displaying and analyzing data

Students use data averaged over several trials to construct a graph corresponding to each concentration of glycerin. Figure 2 provides an example of these plots.

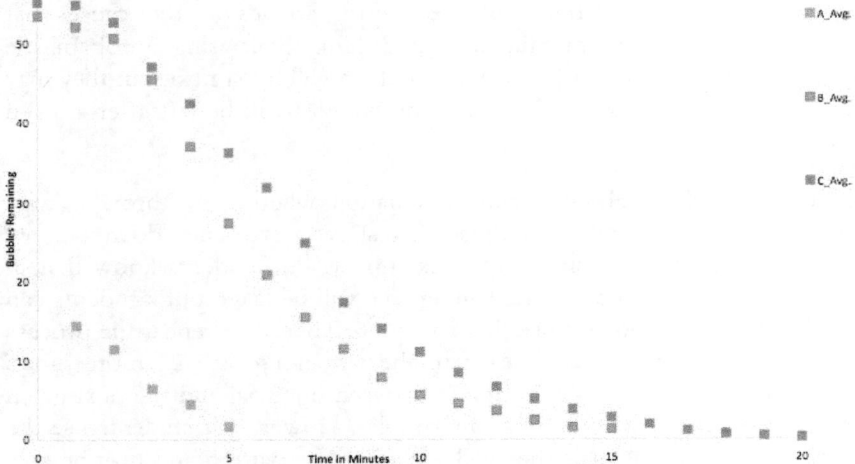

Figure 2. Remaining bubbles for different concentrations of glycerin.

Linear models

Students can visually find a straight line that fits the data fairly well and estimate its slope and intersections with the axes. Then they can use Excel or a graphing calculator to fit a line to the data (Figure 3). Students can discuss to what extent a line with a negative slope is a good model for the number of remaining bubbles. They may notice that in all three cases the linear model predicts zero bubbles earlier than in the actual data. They may want to look for a curve that would fit the pattern of points better.

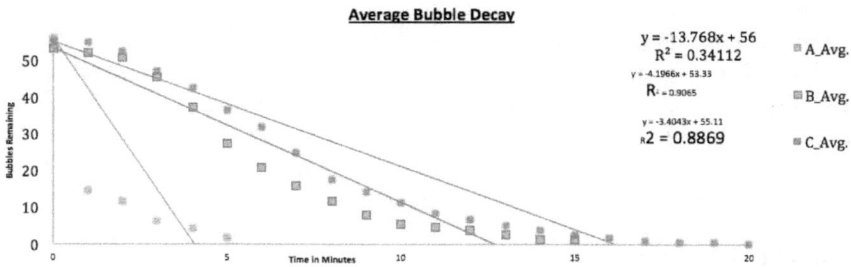

Figure 3. Linear models for different concentrations

Looking for an alternative model

In experiments with very carefully controlled conditions of humidity, temperature, and purity of air (free from dust and carbon dioxide), bubbles can last months (Grosse 1969). In less controlled situations, such as classrooms, bubbles can burst because of dust particles or other causes that can be considered random. Students could think about using a probabilistic model, because they don't know which bubble will burst next, but they may be able predict approximately how many bubbles will be left after a given time.

Students can think about a similar situation where they throw a large number of dice and take out those that show a six. They do not know which dice will show a six on the next throw, but students know that in each throw about 1/6 of the remaining dice will be taken out. Students can toss 56 dice and remove those that show a 6. Then they repeat the process of throwing and removing sixes with the remaining dice 23 more times. There will be some sample variability between one trial and the next given that we are dealing with fairly small samples. However, if students use the averages over several trials they will see a definite pattern and may be able to predict pretty accurately for large numbers of dice the proportion of dice remaining after successive throws. Here are the averages over 11 trials for 24 throws of the dice and the corresponding graph (Figure 4): 56, 47.8, 39.5. 32.5, 28.2, 22.8, 18.6, 15.7, 13.5, 11.7, 9.6, 7.8, 6.5, 5.6, 4.5, 3.8, 3.2, 2.8, 2.4, 2.0, 1.7, 1.4, 1.0, 0.9, 0.6

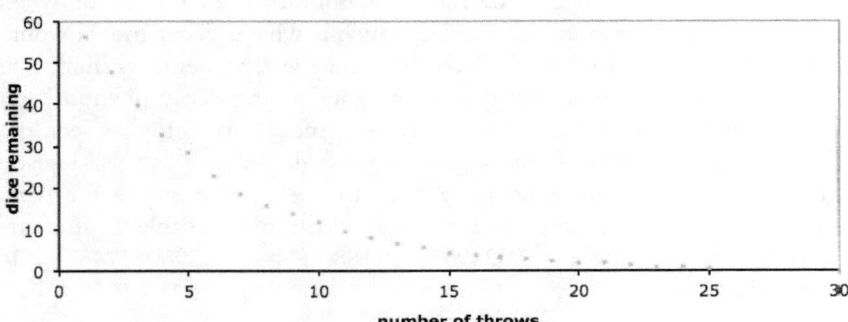

Figure 4. Probabilistic decay data

Use of technology to fit the data

Students can also use technology to fit decay curves to the scatter plots. The graphs in Figure 5 were done using Excel using Trendline feature (exponential function). A template is available from the MEC Lab. Instructions for this template are given in the teacher manual (Hammons 2009b). Students could also use graphing calculators to graph the data and find curves that fit the data.

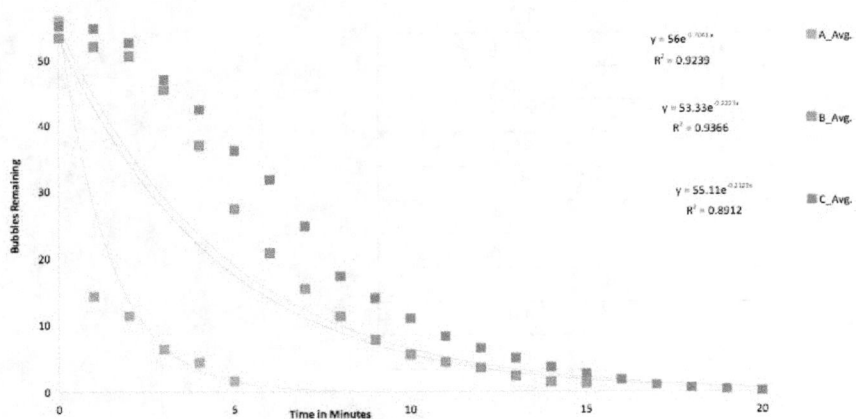

Figure 5. Decay models for bubbles

Students can see that for the bubbles the decay model offers a good fit for the lingering tail, but not quite so for the beginning of the experiment.

Figure 6 shows the graph of the number of bubbles that pop out in each interval of time. The biggest mortality of bubbles does not occur when there is the biggest amount of bubbles, which is what a decay model would predict, but a couple of minutes later. This may lead students to think that the popping of bubbles is not quite analogous to the decay phenomenon with the dice. Students can try a different model to better reflect the behavior of the bubbles. Using a graphing calculator, they can use logistic regression to obtain the equation $y = c/(1+ae^{-bx})$ that best fits the data. Using the data for the averages for 6% concentrations (Table 1) students can obtain the values $a = .3249557511188$, $b = -0.2950237437$, $c = 75.581055453284$

Table 1. Remaing bubbles for 6% glycerin concentration

Time (minutes)	Average number of bubbles remaining
0	55.7
1	53.4
2	47.8
3	43.2
4	37
5	32.7
6	25.7
7	19.1
8	15.8
9	12.8
10	10.1
11	8.33
12	6.78
13	5.44
14	4.44
15	3.56
16	2.78
17	2.33
18	2.11
19	1.89

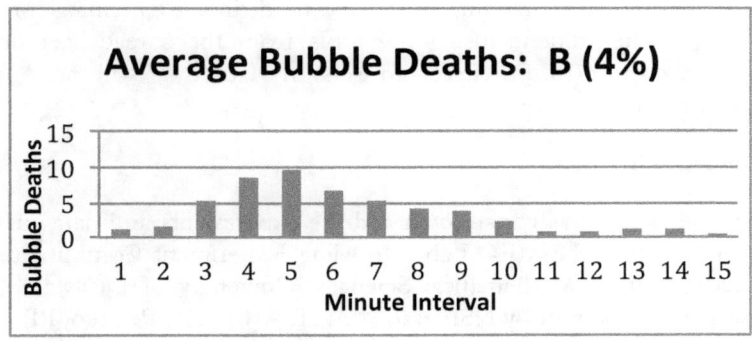

Figure 6. Number of bubbles burst in each time interval (4% mix).

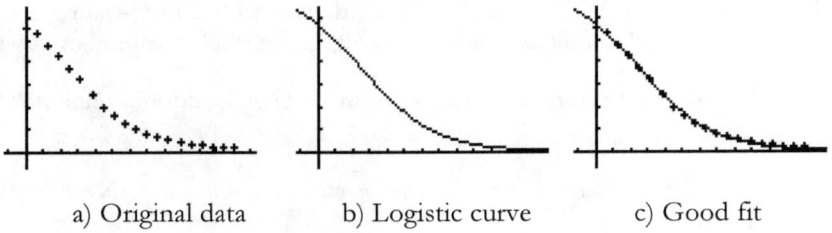

a) Original data   b) Logistic curve   c) Good fit

Figure 7. Logistic model for bubbles

The bubble board in the classroom

There are several ways to use bubble boards in the classroom, ranging from teacher demonstrations to students working in teams and taking turns to collect data. For example, for a class with 24 students and two boards, 12 students would be assigned to each board. They would work in teams of three and take turns to collect data. For the experimental part students in each team need to know what are the necessary tasks to be done (measuring, writing data down, entering data into computer or graphing calculator, etc.) and who is going to be responsible for each task. They can also take turns in the different roles. The activity with the bubble board can be spread out over more than one period. In the first session, students at one board collect data for mix A (2%), about 5 minutes per trial, and mix B (4%) about 10 minutes per trial. Teams at the other board could collect data for mix C (6%) 15 minutes per trial. If more time is available, trials for mixes B and C could be extended to 15 and 20 minutes to better observe the tail. In the following session students bring their data together so that data for the averages of several trials are graphed, rather than individual trials. In following sessions students can then analyze and discuss the graphs and different curves of fit as a group, noting the emergent patterns and considering ways to improve data modeling. Depending on the familiarity of the students using the tools from the spreadsheet or the graphing calculator to fit curves to sets of data they will need more guidance from the teacher.

Acknowledgements

The Bubble Board was constructed and the experiments and data analyses were conducted in the MEC Lab (Modeling Experiment Computation) at the Department of Mathematical Sciences, University of Delaware. This work was funded in part by NSF grant #312154 (John A. Pelesko, PI).

References

Behroozi, F. and D. W. Olson. "Colorful demos with a long-lasting soap bubble." *American Journal of Physics* 62(9): 856-857. September 1994.

Boys, C. V. *Soap bubbles and the forces which mould them*. London: Heinemann, 1959.

Boys, C. V. *Soap bubbles: Their colors and forces which mold them*. New York: Dover, 1958.

Grosse, A. V. "Soap Bubbles: Two Years Old and Sixty Centimeters in Diameter." *Science*, (New Series) 164, no. 3877 (1969): 291-293.

Hammons, Alexandrea Nicole. "Bubble Board." 2009a.
http://meclab.pbworks.com/Bubble-Board

Hammons, Alexandrea Nicole. "Bubble Board Teacher Manual." 2009b.
http://meclab.pbworks.com/Bubble-Board

Isenberg, Cyril. *The science of soap bubbles and soap films*. New York: Dover, 1992.

Rämme, Göran. A method to determine the average lifetime of soap bubbles. *Physics Update* 7, no. 1 (2001): 3-8.

Appendix A

Materials

- Tank, $\geq$ 22 cm deep, 10-14 cm wide, $\geq$ 38 long
- Tray big enough to hold tank (to prevent leakage and spills)
- Ruler, cm

- Straight clear drinking straws
- Permanent marker
- Bubble board
- Large beakers, ≥ 400 ml capacity
- 2.5 cm masking tape (if beakers do not have white label space)
- Pencil and paper
- Scale (grams)
- Small beaker, 100-250 ml capacity
- 12 g dish soap (Dawn)
- 48 g 86-88% Glycerin
- Pipettes
- 1140 ml tap water, plus enough to fill tank
- Stirring rod
- Computer with internet access and Microsoft Office Excel 2007
- 1 – 21.6 by 28 cm (8 ½ by 11 in) Box Frame (sold to frame pictures), or other shallow tray to hold soap solution
- Paper towels
- 56 dice

Appendix B Partial list of providers

Sources for materials and teacher resources

Boards that are perforated can be ordered from Ponoko, http://www.ponoko.com/make-and-sell/how-to-make

Glycerin can be bought in small amount in a pharmacy or in larger quantities, for example at Acros Organics http://www.acros.com/

Straws. A box of 1000 unwrapped clear straws costs a little over $5

The water tank can be purchased from Educational Innovations Inc. Demo tank.

http://www.teachersource.com/BiologyLifeScience/LifeScience/DemoTank.aspx

An on-line stopwatch is available at
Online-Stopwatch. http://www.online-stopwatch.com/split-timer/

Excel worksheet for fitting curves is available from MEC-Lab
http://meclab.pbworks.com/Math-Ed+Team.

10 THE MEANS AS WEIGHTED AVERAGES[10]

Abstract

The quadratic mean, geometric mean, and harmonic mean of two numbers a and b are obtained as weighted averages by using segments parallel to the bases a and b of a trapezoid.

1. Introduction

Often students learn the different means of two numbers, such as the arithmetic average, the geometric mean, the harmonic mean, independently from each other and do not make connections among them, nor do they see why one of the means may be greater than another. Here we provide a unifying view of the different means as weighted averages of the given numbers. We interpret the different means as segments parallel to the bases of a trapezoid (1). The numbers a and b will satisfy $0 < a \leq b$.

2. A geometric interpretation of weighted averages

When computing the arithmetic average and one of the numbers is repeated, say a, a, a, b, we need the weighted average $(3a + b)/4$. The coefficients of a and b are their weights, and the denominator is the total sum of the weights. The segment parallel to the bases a and b of a trapezoid, equidistant from them, represents their arithmetic average. If we take a parallel line segment equidistant from the midparallel segment and one of the bases, its length is the average of the midparallel and the

---

[10] Flores, A. (2011). The means as weighted averages. *Teaching Mathematics and its Applications*, 30(3), 147-150. Reprinted by permission of Oxford University Press.

corresponding base, $(a + (a + b)/2)/2 = (3a + b)/4$ for the upper fourth (Figure 1), or $(a + 3b)/4$ for the lower fourth. Thus the lengths of these segments represent weighted averages of the two numbers. The weight of a base is less if the segment is farther away from this base than from the other.

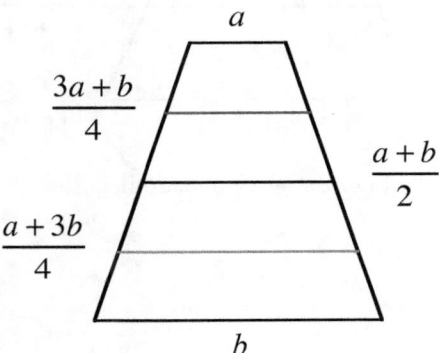

Figure 1. Weighted averages

In general, if the distance of a segment to the base a is n, and the distance to the base b is m (Figure 2), the weight corresponding to a will be m and the weight corresponding to b will be n. The length is given by $\dfrac{ma + nb}{m + n}$ (1) (Prove it!). So, if h_a is the height of the upper trapezoid, and h_b is the height of the lower trapezoid, the weights w_a and w_b associated to a and b satisfy the inverse relation $\dfrac{h_a}{h_b} = \dfrac{w_b}{w_a}$ (2). If we replace m and n by km and kn in expression (1) its value does not change. Thus the corresponding value of x will be the same for any trapezoid with bases a and b, and the same height ratio $\dfrac{h_a}{h_b}$, independently of the height of the original trapezoid.

Given this ratio, the length of the segment can be computed as a weighted average of a and b. The weights for a and b can be any two values that satisfy equation (2).

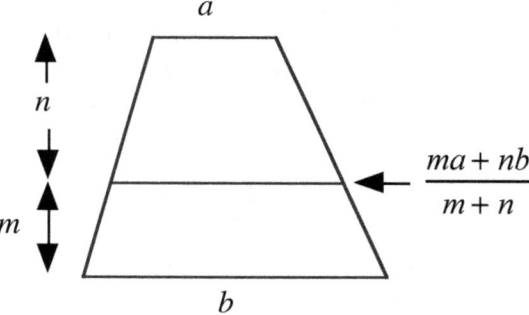

Figure 2. Weights and heights

3. Equal areas

Let's consider the parallel segment that divides the trapezoid in two trapezoids with the same area (Figure 3). The length x of this segment satisfies $h_a \dfrac{a+x}{2} = h_b \dfrac{b+x}{2}$, or equivalently $\dfrac{h_a}{h_b} = \dfrac{b+x}{a+x}$. Thus, the weight of a is $a + x$, the weight of b is $b + x$, and $x = \dfrac{(a+x)a+(b+x)b}{(a+x)+(b+x)}$. Thus $x = \sqrt{\dfrac{a^2+b^2}{2}}$, the quadratic mean of a and b.

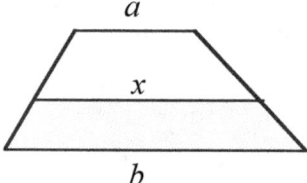

Figure 3. Equal areas

4. Two similar trapezoids

Let's consider the parallel segment that divides the original trapezoid into two trapezoids that are similar to each other. The heights of the two trapezoids are proportional to their larger bases, kx and kb (Figure 4). Segment x is a weighted average where the weight of a is b, and the weight of b is x, and $x = \dfrac{ba + xb}{x + b}$. We obtain $x^2 = ab$, or $x = \sqrt{ab}$. This is the geometric mean of a and b.

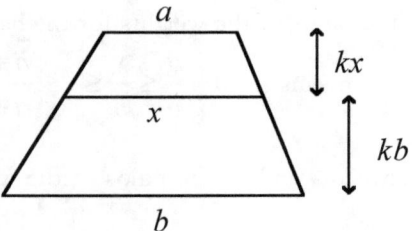

Figure 4. Similar trapezoids.

5. Intersection of diagonals

Let's consider the parallel segment through the intersection of the diagonals of the trapezoid (Figure 5). Triangles ABE and DCE are similar. Their heights are proportional to the bases of the trapezoid. The weight of b is a, and the weight of a is b. The length of the segment is $(a \times b + b \times a)/(a + b) = 2ab/(a+b)$. This is the harmonic mean of a and b. This mean is obtained as a weighted average also when computing the average speed of traveling the same distance d once at speed a and then at speed b,

$$\frac{2d}{\dfrac{d}{a} + \dfrac{d}{b}} = \frac{2ab}{a+b} = \frac{ab + ba}{a+b}.$$

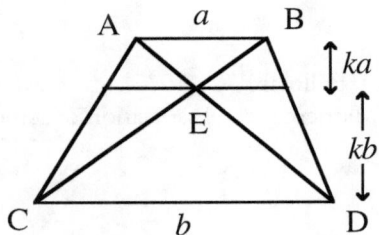

Figure 5. Segment through intersection of diagonals.

6. Concluding remarks

By placing all segments on the same trapezoid (Figure 6) we can see the order of the different means $a \leq \dfrac{2ab}{a+b} \leq \sqrt{ab} \leq \dfrac{a+b}{2} \leq \sqrt{\dfrac{a^2+b^2}{2}} \leq b$. By thinking in terms of the weights we can have additional understanding for this order. For $a \leq b$, the ratios of the weights for the harmonic, geometric, arithmetic, and quadratic means satisfy $\dfrac{a}{b} \leq \dfrac{x}{b} \leq 1 \leq \dfrac{b+y}{a+y}$, where both x and y are numbers between a and b. The ratios of the weights $\dfrac{w_b}{w_a}$ give us additional insight about the order of the different means.

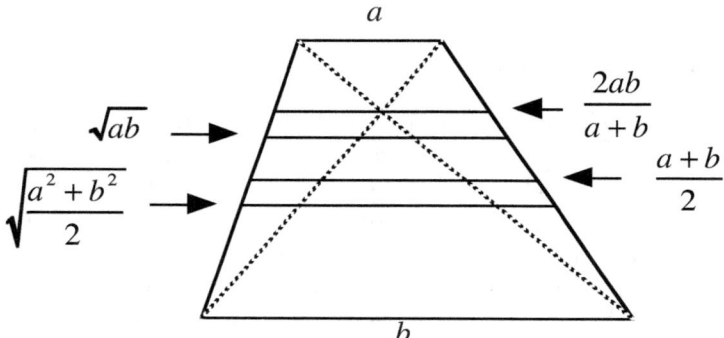

Figure 6. Different means

Reference

1. Beckenbach, E. and Bellman, R. (1975). *An introduction to inequalities.* Washington, DC: Mathematical Association of America.

11 EXPERIMENTS ON THE SURFACE AREA OF THE SPHERE[11]

This article presents an approach where students in middle school or early high school can derive the formula for the surface area for the sphere based on two empirically observed facts. Flexible bendable rods (Wikki Stix) are used to cover the sphere and to cover shapes of known surface area. The first fact students will observe is that the surface area of a whole sphere is equal to the lateral area of a cylinder with base radius equal to the radius of the sphere and with height equal to the diameter of the sphere. The second fact students will observe is that the surface area of half a sphere is equal to the area of two great circles, that is, circles with the same radius as the sphere. These activities can be done by students working individually or in small groups.

Materials
A cylinder, a sphere, and a half sphere are needed. The dimensions of the sphere, the half sphere, and the cylinder are related to each other as indicated. The sphere and the half sphere have the same radius (see Figure 1). The base of the cylinder has the same radius as the sphere and the height of the cylinder is equal to the diameter of the sphere (see Figure 2).

Wikki Stix are bendable rods that are quite strong, made with special yarn covered with a non-toxic wax. They are 8 inches long and stick to any smooth surface. The amount of rods needed will depend on the size of the

---

[11] Flores, A. (2011). Experiments on the surface area of the sphere. *Teaching Mathematics in the Middle School, 17*(4), 203-205. Copyright National Council of Teachers of Mathematics. Used by permission.

sphere. For a sphere of diameter 3 inches, about 36 sticks are needed. Prerequisite knowledge.

Formula for the area of a circle with radius r, $A_c = \pi r^2$. Formula for the lateral surface area of a cylinder given the radius of the base r and the height h, $A_L = 2\pi rh$.

Figure 1. Sphere and half sphere with the same radius.

Figure 2. Height of cylinder equal to diameter of sphere, same radius.

Activity 1. Sphere and cylinder

The student covers up the sphere by winding Wikki Stix around it, starting at the equator and working up to the poles (see Figure 3). In theory, the same number of Wikki Stix should be used to cover each half. In practice, however, there may be some small experimental error due to tiny gaps between the rods or some squishing of the rods. Some adjustments by partially unwinding and rewinding again may be necessary. If the amount used is not exactly the same, students can compute the percentage of error.

Students record how many total rods they used to cover the sphere. For example, in figure 3, six sticks were used for each colored band for a total of 36 sticks. Then the student covers up the lateral surface area of the cylinder by winding the Wikki Stix around it (see Figure 3). It is better to start at the middle of the cylinder and work up to the top and base. Students will notice that they use the same number or almost the same number of Wikki Stix to cover the lateral area of the cylinder as they did for the sphere. Students record the discrepancy, if any, and compute what percentage the discrepancy represents compared to the total surface area of the sphere. Students can compute the percentage error by comparing the difference to the total number of Wikki Stix pieces used. So for example, if the student cannot fit the last 2 inches of a stick, that is, ¼ of a stick, onto the lateral surface of the cylinder and 36 total sticks were used to cover the sphere, the percentage error is 0.25 ÷ 36 which is less than 1%.

Figure 3. The sphere and the lateral surface of the cylinder covered with the same amount of Wikki Stix

By disregarding the small experimental error students can realize that the same number of Wikki Stix was used to cover the lateral surface area of the cylinder and the surface of the sphere. For example, in Figure 3 each color band consists of six Wikki Stix for both the cylinder and the sphere. Students can next describe this equality of number of Wikki Stix in terms of a relation between areas. Students can state that the surface area of the sphere is equal to the area of the lateral surface of the cylinder. Then the students can write an algebraic expression for the lateral area of the cylinder. Students can use their previous knowledge that the lateral area of the cylinder is equal to the area of a rectangle with length equal to the circumference of the basis circle of the cylinder and width equal to the height of the cylinder. The height of the cylinder is equal to the diameter of the sphere, that is twice the radius. So in this case, the circumference of the base is $2\pi r$ and the height of the cylinder is $2r$. Therefore, the lateral area of the cylinder is $4\pi r^2$. Students should conclude that the surface area of the sphere of radius r is thus $4\pi r^2$.

Activity 2. Sphere and great circles

Using the half sphere, the student traces four great circles on paper. Then the student covers up the half sphere with Wikki Stix (Figure 4) and with the same number of Wikki Stix covers two great circles on paper. In this example, the half sphere was covered by six blue, six green and six yellow sticks. To cover the circles it is better to start at the rim and work towards the center of the circles. One circle is covered by six blue and three green sticks and the other circle is covered by three green and six yellow sticks. Some adjustments may be necessary for the fit to be exact. Alternatively, if the fit is not quite exact, the student can record how much of a stick was missing or not used, and compute the percentage amount of experimental error.

Figure 4. Two great circles of the sphere covered with the same number of Wikki Stix as half the sphere.

Disregarding the small experimental error, students can thus verify that the number of Wikki Stix covering half the sphere is equal to the number of Wikki Stix covering the two great circles. Students can used their previous knowledge that the area of each circle is πr^2, so the area of half the sphere is thus $2\pi r^2$, and the area of the whole sphere is therefore $4\pi r^2$, same formula as obtained by the method in the first activity.

Final remarks

Often, students in middle school or early high school are required to memorize the formula for the surface of the sphere without having elements to realize why the formula is true. While students in higher grades may have more advanced mathematical tools, such as calculus, that can help them derive and understand the formula better, for students in the middle school an empirical validation is within their reach and may be quite convincing, and that way they do not just have to rely on an outside authority. Very often, students in the middle school belief a formula is true based solely on the authority of the teacher or the textbook (Flores 2006). Of course, empirical approaches to find the surface area of the sphere are not new. Some authors suggest activities where students peel an orange and cover four corresponding great circles (see for example Lawrence and Hennesy, 2007). One advantage of using Wikki Stix is that they provide a better fit, and students can estimate the percentage of experimental error.

By providing students with approaches where they can either give a mathematical argument for a fact, or at least be able to verify it empirically, we help them expand their repertoire of justification schemes. The teacher can help students shift the authority in establishing truth in mathematics from an outside source to themselves. In addition, by providing opportunities for students in the middle grades to develop short mathematical deductions based on previously accepted facts, they can develop their ability for such deductions, and be better prepared for more advanced courses where the mathematical argumentation becomes more and more deductive, and deductions often comprise several steps.

Acknowledgement

Photographs by Kay Biondi. Used by permission.

References

Flores, Alfinio. "How Do Students Know What They Learn in Middle School Mathematics Is True?" *School Science and Mathematics* 106, no. 3 (2006): 124-32.

Lawrence, Ann, and Hennessy, Charlie. *Sizing up measurement: Activities for grades 6—8 classrooms.* Sausalito, CA: Math Solutions Publications, 2007.

Materials

Cylinders and spheres with matching dimensions are available commercially. These are two examples.
Set of 10 geometric shapes. Learning Resources Inc. Vernon Hills, IL.
http://www.enasco.com/product/TB18004T
GeoModel® Jumbo Relational Solids - 10cm - Set of 17
http://www.eaieducation.com/ProductInfo.aspx?productid=532359
Wikki Stix are available in arts and crafts stores.
http://www.wikkistix.com/

12 STUDENT MADE GAMES TO LEARN THE HISTORY OF MATHEMATICS[12]

Abstract: Prospective secondary mathematics teachers design their own adaptations of popular board and computer games to learn the history of mathematics.

Keywords: Mathematics games, history of mathematics, meaningful practice, motivation, alternative assessment.

INTRODUCTION

There is long experience of using games effectively to learn mathematics at the high school level [5]. Games have also been used to learn mathematics at the college level [2]. Research has documented the effectiveness of games in the learning of mathematics [1, 6]. In most instances games are designed by the teacher, but students can also use their own creativity to adapt existing games or invent new games to learn mathematics [3]. In the same way, students can adapt or invent games to learn the history of mathematics.

In this article we describe how college students, most of whom are prospective secondary mathematics teachers, designed and used their own games to learn the history of mathematics. We begin the article by describing some of the games students designed and used, and follow this

[12] Huntley, M. A. and Flores, A. (2011). Student made games to learn the history of mathematics. *PRIMUS*, 21(6), 1-10. Reprinted by permission of Taylor & Francis (http://www.tandfonline.com).

with a discussion of factors for successful use of games in learning mathematics. We conclude with some of our own reflections, and a reflection from a student.

GAMES DESIGNED BY STUDENTS

The history of mathematics course at the University of Delaware is specially designed to prepare undergraduate secondary mathematics education majors to become better mathematics teachers by enhancing their knowledge of mathematical content and ways of thinking about this content vis-à-vis their future teaching [4]. In this course each student takes lead responsibility for one of the mathematical topics explored throughout the semester. Through their assigned readings, writing assignments, and in-class presentations to their peers, students learn about the historical aspects of all the topics, and investigate deeply the historical aspects of their assigned mathematical topic. One of their assignments involves preparing and delivering a 45-minute mathematics lesson on an aspect of their topic. This lesson they present to their classmates. Some of the students choose to use games as a mechanism for incorporating the history of mathematics into their lessons. After their presentation students write a 3-4 page paper summarizing their lesson. Their paper includes reflections on how they perceived their lesson played out.

The games that students in the history of mathematics course used in their lessons during Spring 2008 and Spring 2009 took on a variety of forms. Several students simply posed problems the quickest. Rewards generally consisted of a piece of candy or verbal praise. Other students designed games that are adaptations of popular commercially-available board games or television game shows. The kind of questions that students chose is what made them appropriate for the history of mathematics class.

Most students who used a game in their lesson incorporated this at the end of their lesson. These students began their lesson by reviewing key concepts from their prior presentation, including the historical and cultural influences on the development of their mathematical topic, and the life of people who played a major role in developing the mathematics. After this review the presenter generally provided some problems to practice the content covered, and then played a game at the end of class, if time allowed. One exception was a student who used a game as a warm-up activity during his lesson.

In the sections that follow we provide several examples of students using games in their lessons. These students were enrolled in the Spring 2008 or Spring 2009 offering of the history of mathematics course at the University of Delaware.

Example 1. Laurel's assigned mathematical topic was fractions. As part of the lesson she presented to class, she engaged her peers in a game of BINGO. Laurel downloaded from the Internet the template for the BINGO cards (http://www.education-world.com/a_lesson/TM/WS_lp268-04.html). She wrote "answers" in each blank space of the grid (see Figure 1).

$\frac{5}{28}$	$14\frac{1}{9}$	.002995	$\frac{8}{12}$	$\frac{37}{5}$
6	64	$\frac{106}{9}$	$\frac{10}{50}$	$\frac{31}{45}$
$\frac{168}{5}$	$3\frac{7}{8}$	FREE SPACE	$\frac{4}{3}$	$\frac{76}{9}$
$\frac{3}{19}$	$\frac{107}{25}$	3	$\frac{-29}{24}$	$\frac{2}{3}$
$-\frac{7}{15}$	$\frac{11}{19}$	$\frac{52}{25}$	0	.75

Figure 1. Sample BINGO board.

To play the game, Laurel wrote mathematics problems, one at a time, on the chalkboard, reading them aloud as she wrote. Three of the problems are as follows:

(1) $4\frac{2}{3} + 7\frac{1}{9}$

128

(2) Rewrite 31/8 as a mixed number

(3) Write as a simple fraction: $\dfrac{3}{4} \div \dfrac{9}{8} \div \dfrac{1}{2}$

Laurel's peers were to complete each arithmetic problem and place a marker on the corresponding box on their BINGO board to indicate the solution. The first player to get five markers in a row (horizontally, vertically, or diagonally) was the winner. Laurel's written summary of her lesson included the following.

> My goal in playing this game was to relate the topic to a fun game that everyone knows. I want my students to see that math can be fun . . . . Fractions seem like an easy concept that each student, especially in college, should understand inside and out. In my observations at a nearby high school for an education class, I observed the exact opposite. The students were in ninth grade and were struggling with simple operations on fractions. This supports my argument that fractions are important in all grades and need to be reviewed after we learn them in elementary or middle school.

Example 2. Dorothy's assigned mathematical topic was whole numbers, focusing on writing whole numbers in various numeration systems. These systems include the tally system, the Egyptian hieroglyphic numeration system, the Babylonian numeration system (based on two wedge-shaped symbols), the Mayan system (consisting of dots and bars), the Roman numeral system, and the Hindu-Arabic system.

During her lesson, Dorothy divided the class into teams. A representative of each team went to the board and solved a conversion problem. The persons at the board received assistance from their peers, who simultaneously solved the problem at hand at their desks. The first set of problems involved converting to or from the Hindu-Arabic system. For instance, one problem involved converting 1213 to the Egyptian system, and another problem involved converting the numeral represented by the Babylonian symbols ▼▼▼　◀▼　◀◀▼▼▼ to the equivalent Hindu-Arabic numeral. The second set of problems involved converting from one numeral system to another. For instance, students were asked to provide the Babylonian equivalent for the set of symbols given in Mayan notation shown in Figure 2.

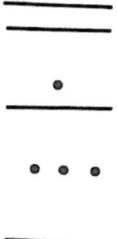

Figure 2. Mayan notation.

Dorothy hung posters on the front board to assist the teams in their conversions. The posters illustrated the various numeration systems. Students were permitted to use calculators during this game, and for each problem, the team who correctly wrote the correct solution to the conversion problem on the board received a piece of candy.

All students actively participated in this game. The people at their desks solved the problems and made suggestions to their teammates at the board, sometimes overly enthusiastically.

Example 3. Michael James's assigned mathematical topic was negative numbers. During his lesson he used an adaptation of a game called *Integer Flash* [7]. For this game students are in groups of three. Each group is provided with a deck of playing cards and a die. Number cards have their face value, and jokers = 0, aces = 1, jacks = 11, queens = 12, kings = 13. Red cards represent negative values and black cards represent positive values. The goal of the game is to be the fastest person to solve integer computation problems. In each group the deck of cards is shuffled and evenly split between the group members. Players keep their cards face down in a pack. The game begins by rolling the die twice. The result of the roll determines the operations to be performed, with 1 = addition, 2 = subtraction, 3 = multiplication, 4 = division. (If a 5 or 6 is rolled, the die is rolled again.) Each player simultaneously flips over one card from her/his pack, and the group members use the cards, together with the operations, to perform the computation clockwise. The first person to correctly complete the computation wins that round. The person who wins the most rounds is the winner. The original version of the game involved students being grouped into pairs, and performing an integer computation involving just two numbers and one operation. In his written paper, Michael James noted that he adapted this game by having students grouped into triads and performing two operations with three cards because "I felt it would be too simple for college students to simply perform one computation with integers."

Example 4. Adaptations of the game Jeopardy were used frequently. The other students enjoyed competing either individually or in teams in such presentations.

4a. Heini was an atypical student in the history of mathematics class. Instead of being a prospective secondary mathematics teacher, he plans to attend medical school upon graduation. His assigned topic was electronic computers. As part of his presentation, Heini engaged the class in game he created called "Computer Jeopardy." Using a computer, he projected the following Jeopardy board onto the screen (see Figure 3).

INVENTORS	INVENTION/ CONTRIBUTION	BINARY ARITHMETIC	RANDOM
100	100	100	100
200	200	200	200
600	600	600	600
800	800	800	800
1000	1000	1000	1000

Figure 3. Jeopardy board.

Students in the class were divided into two teams. Teams took turns selecting a category and dollar amount. Corresponding to each, Heini selected the appropriate cell on the board and the computer program displayed a question. If students supplied the correct answer, that team received the points. For instance, the 200 point question for the category Binary Arithmetic was: *Convert 75 (decimal) to binary.* The category Heini labeled "Random" included questions about his own personal history. This assessed the extent to which his peers had gotten to know him throughout the semester. Questions in this category included the following: *Where am I originally from?* and *What is my favorite baseball team?*

4*b*. Maura used a computer version of Jeopardy done in PowerPoint. Her topic was the history of cubic equations. The categories for her game were The Binomial Theorem, Chinese Cube Root, Indian Cube Root, Conic Sections, and Algebra. Before playing the game, when she presented her topic in class, she illustrated the Chinese method with the image of a cube broken into eight parts (see Figure 4).

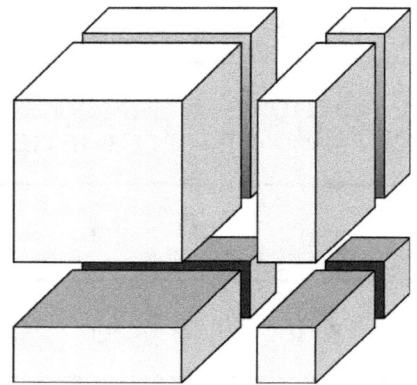

Figure 4. Decomposition of $(r + s)^3$.

During the Jeopardy game, two of the questions in the Chinese Cube Root category were related to this figure: *The binomial expansion breaks apart this shape into 8 pieces,* and *Before being broken up, what is the side of the cube equal to?*

Some of Maura's questions concerned historical facts, like *Who published the book Arybhatiya?* and *This book is where the Chinese Cube Root Method is found.* For the Final Jeopardy question Maura asked students to solve the cubic equation $x^3 + 9x - 26 = 0$ using Cardano's formula.

4*c*. Monica's topic was the history of linear equations. The categories for her Jeopardy game were People, Places and Things; The Method of False Position; The Method of Double False Position; and The Method of Al-jabar. Some of her questions were about facts, such as

(1) *The Method of False Position was first developed in this country, as seen in the Rhind Papyrus* (What is Egypt?); and (2) *The name of the influential mathematician who wrote "al- jabar"* (Who is al-Khwarizmi?).

Some of Monica's other questions required some understanding of the different methods used by mathematicians of different ages and places to solve linear equations. Two examples are as follows: (1) *The first step in solving an equation using the Method of False Position* (What is taking a guess?); and (2) *The factor by which one would multiply an original guess of 7 when solving the following equation using the Method of False Position:* $x + (1/7)x = 16$ (What is 2?).

4d. Fanny's topic was writing algebra with symbols. The categories for her Jeopardy game were Symbol to Person, Person to Symbol, Symbolic Expressions, and Miscellaneous. Her questions were mainly about notations used in different ages, such as the following: (1) *How would Thomas Harriot write* $6a^3 + 4b^2 - 5c$? ($6aaa + 4bb -5c$); and (2) *What would this equation be equivalent to today? Cu.p.7.ce.m.3.co − 4.co.p.2 ?* ($x^3 + 7x^2 -3x = 4x +2$).

FACTORS FOR SUCCESSFUL USE OF GAMES

Like any instructional method or material, the success of using games to facilitate students' learning the content depends on how the games are used. The same principles that make games effective in the learning of mathematics [5] are also valid for games used in the learning of the history of mathematics. In this section we discuss how the games described above satisfy five such principles.

First, each of the games addresses the needs of the history of mathematics class. The games contribute to students' learning history of mathematics facts and ideas in ways that would be hard to attain better using other approaches. The history of mathematics is replete with names, facts, and notations that are not easy to remember. By playing the games illustrated above, students practice the names, facts, and notations in a setting they enjoy, rather than trying to memorize by rote. All the games illustrated above involve facts and concepts related to the course, and their main emphasis is to help students become familiar with ideas that are important to understand the development of mathematical ideas through the ages and in different places. All students did participate actively in the games. Each of the games introduced an element of chance to give everybody (not just the more able students) the opportunity to win and be successful.

Second, the games were used at appropriate times in the course, during a regular class session when the ideas or facts were being taught. Games were relatively short so that students did not lose interest.

Third, the games were planned carefully and conducted in an organized way. The informality and excitement did not interfere with the purpose of the game. Before the game students were instructed on the purpose of the game, its rules, and the way to play it. The presenters established ground rules so that everybody enjoyed the game. Usually the loss of points for breaking rules or getting too loud and excited is enough to keep appropriate behavior.

Fourth, students accepted responsibility for learning from the game. We discussed briefly and shared comments after each game. In other settings follow-up activities such as additional readings could be used to emphasize this responsibility. The instructors also evaluated informally to what extent the game was successful in promoting the learning desired. It was interesting to see how much specific information students were able to retrieve from their readings and presentations by their fellow students.

Fifth, each game had a title, rules, and instructions of how to play it. Additionally, all the necessary materials were included. Their size was such that they were easily used (commercially available board games are a good practical reference), and were made of sturdy materials with an attractive appearance. Some of the students also took advantage of available computer resources to develop their games. In this case, the computer program was easy to use and reliable.

FINAL COMMENTS

In the history of mathematics course offerings during Spring 2008 and Spring 2009, students actively and enthusiastically participated in the games during class. They made their best effort to get the right answers and to do well in the games. To make the game more fun, students sometimes included in their games small rewards for correct answers such as a small piece of candy, or slightly embarrassing punishments for wrong answers such as being voted off their team. Several mechanisms were used to determine assignment of students into groups for playing the games (e.g., counting off). At the beginning of play, sometimes students would play another game (e.g., rock-paper-scissors) to determine the order of play.

The design and use of instructional materials and educational games had several benefits for the majority of the students. Students needed to apply their knowledge of history of mathematics in the design of their games. Students felt motivated and interested in the history of mathematics and saw how they could integrate it into their own teaching of mathematics. Lastly, they felt better prepared for the new world that will be teaching in their own classrooms.

These ideas are summarized by Rose, who explained her rationale for designing a game and infusing competition into its execution, all in the context of her future teaching.

Because of the difficulty of the [quadratic problems phrased in historical language], I had the students work in groups, so they could bounce ideas off one another. In this way they could learn from each other, because what one student doesn't understand, another one may. This would not just be group work though, but also a competition between groups. Each group received the same problem, and the first three groups to get an answer would choose a member to put their work on the board. As a class, we would compare answers and strategies, and debate over which answer is correct. I chose to introduce the problems this way, instead of doing an example, and then have them copy it down, because I have learned from one of my education courses that giving students a problem they have never seen before, and having them grapple with it allows them to think critically, makes them pull from prior knowledge, and try to transfer that knowledge into a new situation. The competition strategy gives the students incentive to want to solve the problem, which I feel is a great way to engage the students.

REFERENCES

1. G. W. Bright, J. H. Harvey, and M. M. Wheeler. *Learning and mathematics games.* National Council of Teachers of Mathematics, Reston, VA, 1985.

2. S. Forman and S. Forman, *Mathingo: Reviewing calculus with bingo games*, PRIMUS 18 (2008), pp. 304-308.

3. I. Gallegos and A. Flores, *Student-made games to learn mathematics.* PRIMUS (in press)

4. M. A. Huntley and A. Flores, *A history of mathematics course to develop secondary mathematics teachers' knowledge for teaching.* PRIMUS (in press).

5. Johnson, D. A. and G. R. Rising, *Guidelines for teaching mathematics* (second ed.). Wadsworth Publishing Co, Belmont, CA, 1972.

6. J. M. Randel, B. A. Morris, C.D. Wetzel, and B. V. Whitehill, *The effectiveness of games for educational purposes: A review of recent literature*, Simulation and Gaming 23 (1992), pp. 261- 276.

7. J. Beery, G. Cochell, C. Dolezal, A. Sauk, and L. Shuey. 2005. Negative numbers. In Katz,V. J. and K. D. Michalowicz. (Eds.), *Historical modules for the teaching and learning of mathematics [CD-ROM].* Washington, DC: Mathematical Association of America.

ACKNOWLEDGEMENTS

The authors express appreciation to Dr. Kathy Clark, Florida State University, for helping to conceptualize the development of the history of mathematics course at the University of Delaware during Spring 2008 and 2009.

13 BUILDING A PROFESSIONAL KNOWLEDGE BASE FROM RESEARCH AND PRACTICE[13]

Introduction

Mathematics educators often find themselves trying to strike a balance between theory and research on one hand, and the practice of solving concrete and specific problems in the learning and teaching of mathematics or the preparation of teachers of mathematics on the other. Mathematics educators sometimes use theory and research to illuminate their practice, and they sometimes use the problems derived from practice to guide their research. Frequently mathematics educators strive to contribute to a professional knowledge base both from the knowledge they generate through their research as well as the knowledge they generate through their practice.

In this presentation I discuss some characteristics of knowledge for a professional knowledge base. Then I illustrate how knowledge derived from research has those characteristics. Next, I discuss some ways in which research can benefit practice, the ways the field has tried to facilitate the contribution from research to practice, and the difficulties of translating research into practice. The discussion then shifts to highlight the value and importance of practitioner knowledge, and discuss some of its limitations and ways to overcome such limitations. After this, I discuss some features of knowledge building systems. At the end I discuss how a new journal

---

[13] Flores, A. (2011). Building a professional knowledge base from research and practice. In M. B. Swarthout and D. L. Jones (Eds.) *Proceedings Third and Fourth Conferences on Mathematics Education Research in Texas* (pp. 17-27): Sam Houston State University, Huntsville TX. Reprinted by permission of MERiT.

could be one of the mechanisms necessary to develop a professional knowledge base from practitioner knowledge.

How research knowledge contributes to a professional knowledge base

Individuals generate knowledge from their practice and from research, but not all knowledge generated by individuals becomes part of the professional knowledge base. That is, individual knowledge does not automatically become part of the profession's knowledge. Hiebert, Gallimore, and Stigler (2002) identified several characteristics that knowledge generated in a field needs to have be able to contribute to a professional knowledge base. According to them, to contribute to a professional knowledge base knowledge needs to be public, shared, stored, verified, and improved over time.

In this section I describe how knowledge generated from research in mathematics education has the characteristics mentioned above and how there are institutional mechanisms in place and available for researchers so that the research knowledge generated by individuals does indeed becomes part of a professional knowledge base. Research in mathematics education is made public, and shared through presentations at meetings such as PME and PME-NA, the Research Pre-session of NCTM, and AERA's SIG in Mathematics Education sessions. Research knowledge is also shared and stored in publications such as conference proceedings, monographs, and books. The research knowledge is made public and shared, and also stored in journals specific to the field such as *Educational Studies in Mathematics*, the *Journal for Research in Mathematics Education*, and *Mathematical Thinking and Learning*, as well as in other professional journals in education and related fields.

Research knowledge is verified in several ways. There is a well-established tradition of peer review for research articles. For example, each of the research journals mentioned above has a refereed process in place. Articles are carefully read and criticized by other experts in the field. Also, there is a well-established practice that researchers need to back their claims by evidence, and make clear to their readers to what extent the claims they make are supported by their data.

Research journals also strive for an improvement over time in the research they publish, although they may use a different language to describe this quest for improvement. Instead of using the phrase "improvement over time" the phrases "move forward" or "advance the field" are used in JRME. For example, in an editorial, NCTM's Research Advisory

Committee (2001) described how it was "involved in five initiatives intended to help us move forward in making sense of the changing nature of mathematics teaching and learning in our schools." (p. 445) Recently, JRME's editor characterized the kind of articles the journal is looking for. "Research that advances the field of mathematics education ... recognizes and builds on related research and is situated in a way that clarifies its unique contribution." (Heid 2010, p. 436). A well-established practice in research is that authors base their work on previous research and theories, and make clear to what extent does their contribution expand or modify previous knowledge.

Universities value highly that the research knowledge generated by their faculty is made public and shared. Criteria for promotion and tenure at many universities count heavily the publication record of candidates. Universities also value that knowledge is verified by giving more weight to publications in peer reviewed journals. In most research universities a solid record of research publications is necessary to attain tenure and be promoted.

Impact of research on practice

Research in mathematics education can also contribute to a knowledge base that goes beyond what is interesting from a research perspective. Silver (2003) described three ways in which research can contribute to practice. The first is that theoretical perspectives can be useful in framing and describing practical issues and problems. The second is that research methods can illustrate interventions and data-collection processes with practical utility. And the third is findings that can be generalized and used in applied settings. Blume (2010) discussed how "tasks used in research settings can be valuable ... if they offer rich sites for learning mathematics or if students' approaches to them provide useful indicators of their conceptual understanding." (p. 1) Researchers in mathematics educators have for several decades now tried to build bridges from research to practice. Suydam and Dessart published two books on *Classroom ideas from research* in which they extract from many research studies ideas that might prove useful to school practitioners (Suydam and Dessart 1976, Dessart and Suydam 1983). The book series *Research ideas for the classroom* (Wagner 1993, Jensen 1993, Owens 1993, Wilson 1993) provides a practitioner friendly summary of research on different aspects of learning and teaching mathematics. In *Lessons Learned from Research*, Sowder and Schappelle (2002) present a collection of articles for practitioners each of which is based on a research article published in JRME. Each of NCTM's school journals for a number of years has published a department that strives to bring research into practice, such as Research, Reflection, Practice a department in *Teaching*

Children Mathematics; Informing Practice a department in *Mathematics Teaching in the Middle School*; and Connecting Research to Teaching a department in the *Mathematics Teacher*.

Translating research into practice is not easy. Furthermore, not all research is conducted with an eye to have a practical impact. Sowder (2000) lists three necessary conditions for research to have an impact on practice. The first is that our research must be sufficiently persuasive and authoritative before it can affect practice. The second is that research must be relevant to practice. And the third is that ideas from research must be accessible to teachers. The reality however, is that little of the research in mathematics education is ever translated into practice. Sowder points out that the difficulties of affecting practice may include the education system itself which at times appears intractable and unable to change, or, conversely, appears inherently unstable, overly susceptive to fads, and consequently unable to engage in systematic change. Hiebert, Gallimore, and Stigler (2002) discuss also the intrinsic difficulties of translating traditional research knowledge in education into forms that teachers can use to improve their practice.

Building a professional knowledge base from practice

We can also think about generating knowledge for a professional knowledge base in a radically different way. Rather that relying only on knowledge generated by research, we can think on how we can tap on the knowledge generated by practice to develop such a professional knowledge base. I first came across this idea when reviewing a grant proposal (Choike, Stigler, Kornstein, Johanek, Novak, and Suarez, 2002) in which it discussed contributing to a professional knowledge base from teachers' practitioner knowledge. Similarly, we can also think about the teaching practice of professors at institutions of higher education. Shulman (2004b, p. 455) proposes three strategies by which to make teaching, an important aspect of our practice, community property and thus more valued in institutions of higher education. The first strategy to make teaching community property, is to recognize that the communities are identified with the disciplines of our scholarship, including the process of exercising "quality, control, judgment, evaluation, and paradigmatic definition." (p. 456). The second strategy is that teaching "must be made visible through artifacts that capture its richness and complexity." (p. 457). The third strategy is that peer review must be applied to teaching. The artifacts of teaching must be made available so they can be evaluated by communities of peers at other institutions. Hiebert, Gallimore, and Stigler (2002) outline how we can approach the issue of developing a professional knowledge base by starting with practitioner knowledge.

The value and importance of practitioner knowledge

The value of practitioner knowledge has long been recognized in the field of mathematics education. Davis (1967) even stated, "The best practice of the best practitioners almost certainly lies ahead of the best theory of the best theorists." Since then, several authors have stressed the importance of identifying and learning from best practice. Shulman has highlighted the importance of the wisdom of practice in education and pointed to the necessity "to study accomplished practice as it actually occurs and to ask how it has been achieved." (Shulman, 2004c p. 252).

Practitioner knowledge of mathematics teacher educators has features that make it valuable and provide the potential of directly impacting and improving the profession. Hiebert, Gallimore, and Stigler (2002) discuss three features that make practitioner knowledge especially relevant for the profession. The ideas in this section are mainly adapted from that article.

Practitioner knowledge is useful for practice because this knowledge addresses specific problems of practice. The knowledge generated by a mathematics teacher education practitioner is relevant for another mathematics teacher educator addressing the same problem. In mathematics teacher education, practitioner knowledge is linked with practice in two ways. First, its creation originates in problems of practice. Second, the new knowledge is connected to the actual process of preparing teachers of mathematics.

Practitioner knowledge is concrete, specific and detailed. The knowledge generated when solving a specific problem in mathematics teacher education provides someone else facing the same problem with concrete, specific, and detailed knowledge that can be used to solve the problem. Even though practitioner knowledge may not be as generalizable as knowledge generated by research, it is still useful because of these characteristics.

Practitioner knowledge is integrated and organized around complex and multifaceted problems of practice. In contrast to researchers who often try to separate different factors or focus on only certain aspects of a problem, practitioners use all kinds of knowledge that can help them address a problem. For example, when preparing teachers of mathematics to help middle school students understand the concept of function, a mathematics teacher educator would use content knowledge, pedagogical knowledge, and pedagogical content knowledge, as well as their knowledge of the students.

Overcoming limitations of practitioner knowledge

On the other hand, practitioner knowledge has some limitations that in the past have prevented us to build systematically a professional knowledge base from practitioner knowledge. Practitioner knowledge is personal and often not subjected to thorough examination and evaluation. We can get some sense of the difficulties of learning from practice, that is, from own experience, by looking at how hard it is to learn to become a better teacher by reflecting and learning on our own teaching.

Shulman (2004a) states that "learning to teach is enormously complex, a clear example of learning the highest-order forms of knowledge, skill, and problem solving" (p. 319-320), and that "the more complex and higher-order the learning, the more it depends on reflection—looking back—and collaboration—working with others." (p. 319). Shulman's case is that in order for us to learn from experience it has to be a collegial endeavor rather than an individual one.

To make learning from practice a collegial activity and overcome the limitations of practitioner knowledge, we can use Hiebert, Gallimore, and Stigler's framework. In order to contribute to a professional knowledge base, practitioner knowledge in mathematics education needs to be public, stored, shared, verified, and improved over time. We need to change the way we view at our practice from and individual and private endeavor to a scholarly activity that is public and subject to peer review. In addition, to have progress we need not just innovations, we need evidence that the changes are indeed improvements. Furthermore, we need to show the effectiveness of the changes beyond the initial innovators (Moore, 1995).

Features of knowledge-building systems

A special issue of the *Elementary School Journal* edited by Hiebert and Morris (2009), describes and illustrates features of knowledge-building systems. These features include shared goals, tangible products, small tests of small changes, and multiple sources of innovation. Morris and Hiebert (2009) describe and give examples of how "professions that self-consciously improve view the process of human collaboration to yield accumulating knowledge as an essential mechanism for improving their practices" (p. 430)

Shared goals across the system. Shared goals are clearly articulated and accepted by all members of the profession. "There are immediate, daily, challenging problems faced by everyone in the system whose resolutions improve everyone's work lives. Solving these problems can be thought of as shared goals across the system." (Morris and Hiebert 2009, p. 434).

Visible, tangible, changeable products. Tangible products or visible processes that embody the goals of the system are jointly constructed and steadily improved by incorporating the growing knowledge of the system. "Solutions to the problems are represented as visible, tangible products that are continually refined and improved. The products represent the best solutions to the daily problems that members face and are in a form that members can change…. They can be shared and tested throughout the system." (p. 434)

Small tests of small changes. The professional products and processes, and the information generated to produce them, are vetted for quality and usefulness by collecting just enough data about small-scale changes. Quick feedback and faster accumulation of improvements keep people motivated to improve and solve problems. The frequent collection of small amounts of data—just enough data to tell whether the change has promise decrease resistance to change. (p. 435) Small tests of small changes to an existing product may be more acceptable to those who have worked on the product Replications in other contexts can provide sufficient data to determine whether the change should be incorporated into the system. The aim of system improvements is to raise the effectiveness of the standard or most common practice throughout the profession. Improving the performance of a local workgroup, even spectacularly so, does not yield system-wide improvement. Improvements that make a difference come from raising the standard of practice for the profession as a whole.

Multiple sources of innovation from throughout the system. All members contribute to the construction of products or processes based on their special expertise, ensuring multiple sources of innovation from throughout the system. Everyone is viewed as potential source of ideas. Users generate knowledge, not only use it. Status differential between researchers, designers, and users is minimized. Local experimentation and adaptation is encouraged. Better solutions are created than individuals working alone.

Systematic improvement of mathematics teacher education

Based on the previous discussion we can see that there are two ways to make the improvement of mathematics teaching or mathematics teacher education more systematic. The first is building on previous work. This entails different aspects. To the extent possible, what we publish based on our practice needs to be consistent with related research findings, and have a connection to the knowledge base. It should be grounded on theory or on previously published articles. Make explicit the specific new contribution. Provide evidence that the approach is an improvement and not just a change, that is, report findings with enough warrants. The second way is to enhance the likelihood that others can build on the contribution of the article. We can maximize this possibility by describing promising practices that are testable and making hypotheses and rationales explicit.

A new journal as a mechanism for developing a professional knowledge base from practitioner knowledge

In the same way that researchers have developed mechanisms, such as peer-reviewed journals that build on previous knowledge, to be able to contribute to a professional knowledge base from research knowledge, we need to develop the corresponding mechanisms for knowledge that stems from practice. To implement the vision of developing a professional knowledge base in mathematics teacher education from practitioner knowledge we need a journal where the knowledge can be made public, shared, stored, verified, and improved over time. In the following section I will describe how such a journal was conceptualized by AMTE task force and the joint NCTM-AMTE task force. This section follows the report and call for editor fairly closely.

Audience

The primary audience of the *Mathematics Teacher Educator* will be practitioners in mathematics teacher education, broadly defined as anyone who contributes to the preparation and professional development of Pre-K—12 preservice and in-service teachers of mathematics. Mathematics teacher educators include mathematics educators, mathematicians, teacher leaders, school district mathematics experts, and others. Members of the *Association of Mathematics Teacher Educators* who will receive the journal as a member benefit. NCTM members may choose this journal as one of their secondary options.

Scholarly profile: A peer reviewed journal

The *Mathematics Teacher Educator* journal will be a scholarly peer reviewed journal for practitioners using a double blind review process for feature articles. For special issues or departments the review may be done by a panel, but the process will still be blind about the author's identity. A good part of the contributors to the journal will be from university professors and peer review is an important criterion for counting publications in the promotion and tenure process.

Publication niche

The *Mathematics Teacher Educator* will occupy its own publication niche. There are no other journals with the same mission, goals, and characteristics.

Difference with practitioner's journals. The journal would be different from other practitioners' journals. The *Mathematics Teacher Educator* would be different from the *Mathematics Teacher* not only because it has a different audience (teacher educators rather than teachers) but also because the *Mathematics Teacher Educator* explicitly seeks to systematically build from practitioner knowledge a professional knowledge base that is not only public, shared, and stored, but also verified and improved over time. This idea is not completely new. There are journals for practitioners in other fields that also publish the kind of articles that can contribute to developing a professional knowledge base from practitioner knowledge (such as *The Reading Teacher*).

Difference with research journals. The *Mathematics Teacher Educator* will also be different from other scholarly journals on mathematics teacher education, because its starting point would be practitioner knowledge rather than research knowledge. There are research journals geared for mathematics teacher educators, such as the *Journal of Mathematics Teacher Education.* Research can inform practice, but the translation is often difficult. Because the nature of the knowledge used by practitioners in teacher education is often of a different kind than the knowledge generated by researchers, there is the need to have a venue that uses practitioner knowledge as the base for developing professional knowledge. Of course, research will have an important role in many articles of the journal, as many ideas for improvement will come from practice informed by research.

A forum for a different kind of article. In addition to providing a space for articles that currently are scattered in many journals, the *Mathematics Teacher Educator* will provide a space for articles that would otherwise not been written or published.

Example of a possible article that at present does not have a proper forum:

> In a mathematics methods course two different approaches to learn to teach proportionality for conceptual understanding are contrasted. In one approach future teachers engage themselves with tasks and hands-on materials, and in the second approach future teachers see and discuss a video of students solving the same problems using the same materials, and analyze students' written work. The activities described are specific; the rationale for and the hypotheses that lead to each of the two approaches are given explicitly; the amount of data is enough to determine how to build on from the experience.

Such an article can contribute to knowledge that is improved over time. The results can inform the practice of other mathematics educators at other institutions who may use the advantages of the two approaches to improve their own course. However, because the focus is very specific and the sample size small, it would not be likely for such an article to be published in a research journal.

The journal will contribute to developing the knowledge mathematics teacher educators need to be effective in preparing teachers. One of the aspects of this knowledge is about ways to develop pre-service and in-service teachers knowledge. An article may thus address the issue of teacher knowledge and provide suggestions to increase the knowledge of mathematics educators about the central issues teachers have in developing their own knowledge and what works to support them.

What counts as evidence? Especially relevant to establishing the identity of the journal will be to determine what counts as evidence, what are the warrants needed for practice-based knowledge. Because it is not a research journal **Mathematics Teacher Educator** should not just try to use the same criteria the profession uses for research articles. On the other hand, the journal should not be a space to share mere opinions on mathematics teacher education. Evidence for claims should be provided. In a discussion of warranted assertability Dewey states that,

> The inferred material has to be checked and tested. The means of testing, required to give an inferential element any claim whatsoever to be *knowledge* instead of conjecture, are the data provided by observation—and only by observation. Moreover… it is necessary that data (provided by observation) be *new*, or different from those which first suggested the inferential element, if they are to have any value with respect to attaining knowledge." (Dewey 1991 p. 173)

A useful question could be, "Does the article provide enough information or insight as to guide us as to what the next steps for improvement could be?"

Types of Articles

The journal will contribute to enhance the communication among mathematicians, mathematics educators, other groups of experts that contribute to the education of teachers of mathematics.

Types of articles appropriate for the journal might include

- innovative materials that substantially impact mathematics teacher education

- accounts of development of exemplary educational programs in mathematics teacher education, both pre-service and in-service
  - example: implementation of a masters programs for middle school or high school teachers that better connects to classroom practice and includes mathematics courses that are supportive of teaching 7-12. Not just the description on paper, but data from the actual implementation.
- accounts of innovative uses of video or computer software for mathematics teacher education
- the mathematics taught to teachers, and ways that mathematics is taught to teachers
- reports of effective ways to develop teachers' profound understanding of the mathematics they teach
- developing mathematical knowledge for teaching
- reports of effective ways to develop teachers' understanding and corresponding skills and know-how of pedagogical issues specific to mathematics
  - example: developing teachers' skills in orchestrating discourse in mathematics
- teaching the history of mathematics to teachers
- reports of effective ways to prepare mathematics teachers to deal with issues such as equity, special needs of students, second language acquisition
- lesson study reports including evolution of the lesson and process of improvement
- specific experiences that collect and use small amounts of data that allow educators to build improvements
- replications in different contexts
- revisiting "old" ideas and established practices from a new vantage point or with new insights
- promising practices that can be tested, with clear rationales (why was this tried?) and clear hypotheses
- good practices and models for the preparation of mathematics teacher educators
- professional development for those who teach teachers of mathematics
- discussion of goals for mathematics teacher education with explicit rationales
- effective ways to help teachers implement reform-based curriculum materials

- reports about development projects that contribute to improving mathematics teacher education
- lessons learned from research; applications of research methodology, theoretical frameworks, or results to the practice of mathematics teacher education
- improved methods and instruments for evaluation of mathematics teacher education
- reviews of scholarly books relevant to mathematics teacher education
- reviews of textbooks that are substantially different from and better than previous textbooks and that are especially relevant for mathematics teacher education

Additional features

The journal will have a web-based repository for scholarly artifacts that supplement articles (video, lesson plans, classroom materials, student work, etc.), providing a way to make these materials public and thus opening the possibility for further improvement by other practitioners. Furthermore, having an electronic journal could be quite revolutionary in terms of how we think about sharing practice in mathematics teacher education. An electronic journal could have live links embedded throughout e-articles, have appendices of datasets or student work samples available at the click of a mouse, have videolinks to classroom episodes (instead of, or in addition to lengthy quotes and transcripts within an article). In short, as McLuhan (1964) pointed out, the medium has also the potential to alter the message in important ways.

Where are we now?

AMTE and NCTM Boards approved to jointly launch a new journal for practitioners to implement the vision described above. Call for an editor was posted. A three year agreement between AMTE and NCTM starting July 2012 was signed. An editor and editorial panel have been appointed. The journal will issue a call for papers in the Spring of 2011.

Concluding remarks

Providing systematic collegiate ways to tap on the wisdom of practice to develop a professional knowledge ways can have profound impact in academic life. Although universities value both teaching and research, the situation for both is quite different in terms of how each activity is at present evaluated at most universities. While there are clear standards and procedures for making researcher knowledge part of a collegial activity, the same is not true about teaching or other aspects of practice. Evaluation of teaching does not at present have the same consistency across universities as does evaluation of research, and in most places teaching is still a very individual activity. Providing mechanisms such as a peer reviewed journal is one important step to make it possible for knowledge that stems from practice to contribute to a professional knowledge base.

References

Blume, G. W. (2010). Research articles can be helpful to practitioners. *Journal for Research in Mathematics Education, 41*(1), 2-5

Choike, J., Stigler, J., Kornstein, S., Johanek, M., Novak, T., Suarez, A. (2002). Preparing, inspiring, and connecting underserved students to college and opportunity. A College Board Proposal to the National Science Foundation. Unpublished manuscript.

Davis, Robert B. (1967). The range of rhetorics, scale, and other variables. *Journal of Research and Development in Education, 1*, 51-74.

Dessart, Donald J. and Suydam, Marilyn N. (1983). *Classroom ideas from research on secondary school mathematics*. Reston, VA: National Council of Teachers of Mathematics.

Flores, A., Arbaugh, F., Lannin, J., McGraw, R., Rubenstein, R., Stallings, L., Wilson, P. 2010. AMTE Journal Task Force. Final report and recommendations *Mathematics Teacher Educator*. Presented to AMTE Board.

Flores, A., Kepner, H., Lambdin, D., Lannin, J., Rubenstein, R., and Sowder, J. Joint AMTE – NCTM Task Force. Call for editor. http://www.nctm.org/news/content.aspx?id=26485

Heid, M. K. (2010). The task of research manuscripts—Advancing the field of mathematics education. *Journal for Research in Mathematics Education, 41*(5), 434-437.

Hiebert, J., & Morris, A. K. (2009). Building a knowledge base for educating (mathematics) teachers [Special issue]. *The Elementary School Journal, 109*(5).

Hiebert, J., Gallimore, R., & Stigler, J. W. 2002. A knowledge base for the teaching profession: What would it look like and how can we get one? *Educational Researcher, 31*(5), 3-15.

Jensen, Robert J. (1993). *Early childhood mathematics: Research ideas for the classroom*. New York: Macmillan.

McLuhan, Marshall. (1964). *Understanding media: The extensions of man*. New York, McGraw-Hill.

Moore, David S. (1995). The craft of teaching. MAA Focus 15 (1995) Number 2, 5-8.

Morris, A. K., & Hiebert, J. (2009). Introduction: Building knowledge-bases and improving systems of practice. *The Elementary School Journal, 109*(5), 429-441.

Owens, Douglas T. (1993). *Middle school mathematics: Research ideas for the classroom*. New York: Macmillan.

Research Advisory Committee. (2001). Supporting communities of inquiry and practice. *Journal for Research in Mathematics Education, 32*(5), 444-447.

Shulman, Lee S. (2004a). Teaching alone, learning together: Needed agendas for the new reform. In Shulman, Lee S. *The Wisdom of Practice: Essays on Teaching, Learning, and Learning to Teach* (p. 309-333). San Francisco, CA: Jossey-Bass.

Shulman, Lee S. (2004b). Teaching as a community property: Putting an end to pedagogical solitude. In Shulman, Lee S. *The Wisdom of Practice: Essays on Teaching, Learning, and Learning to Teach* (p. 455-459). San Francisco, CA: Jossey-Bass.

Shulman, Lee S. (2004c). The wisdom of practice: Managing complexity in medicine and teaching. In Shulman, Lee S. *The Wisdom of Practice: Essays on Teaching, Learning, and Learning to Teach* (p. 251-271). San Francisco, CA: Jossey-Bass.

Silver, Edward A. (2003). Border crossing: Relating research and practice in mathematics education. *Journal for Research in Mathematics Education, 34*(3),182-184.

Sowder, Judith T. (2000). Relating research and practice. Retrieved January 19, 2006, from http://public.sdsu.edu/CRMSE/js_nctm_pre.html

Sowder, Judith T. and Schappelle, Bonnie (Eds.) (2002). *Lessons learned from research*. Reston, VA: National Council of Teachers of Mathematics.

Suydam, Marilyn. (1976). *Classroom ideas from research on computational skills.* Reston, VA: National Council of Teachers of Mathematics.

Wagner, Sigrid (Director). (1993). *Research Interpretation Project.* Reston, VA: National Council of Teachers of Mathematics.

Wilson, Patricia S. (1993). *High School Mathematics: Research ideas for the classroom.* New York: Macmillan.

14 ERDÖS NUMBER 1... FOR MOUNTAIN CLIMBING[14]

The prolific mathematician Paul Erdös is at the center of a system to number mathematicians. Erdös had many coauthors; they are assigned Erdös Number 1. Mathematicians who have coauthored with a mathematician who is Erdös Number 1 are assigned Erdös Number 2, and so on (see Erdös Number Project for details). Erdös is known for several eccentricities, but perhaps it is not as well known within the mathematical community that something Erdös liked to do outside mathematics, as he visited places around the globe, was to take hikes on nearby mountains accompanied by local mathematicians. Thus, a similar system could be devised for mountain climbing. People who climbed a mountain with Erdös would be given Number 1, and so on.

When Erdös visited UNAM in Mexico City around 1978, he requested to be accompanied on a hike to the highest nearby mountain that was not covered with snow. Víctor Neumann, the professor who had invited Erdös, declined to do so himself because of health problems. So, a group of six graduate and undergraduate students was recruited to go with Erdös.

There are many remarkable mountains around Mexico City. Because of Erdös' request of no snow on the mountain, majestic Popocatépetl (5500 m above sea level) and amazing Iztaccihuatl (5286 m) were not considered. We decided to take Erdös to Ajusco (3930 m), a beautiful mountain to the south of the city. It was not quite the highest mountain without snow around the city, but it was closer and more accessible than either Tláloc

---

[14] Flores, A. (2011). Erdös number 1... for mountain climbing. *MAA FOCUS, December 2011 / January 2012,* p. 18. Reprinted by permission.

(4158 m) or Telapón (4085 m). Practical considerations took precedence over strict adherence to Erdös' request for the highest mountain. Of course, we did *not* tell him that Ajusco was not quite the highest mountain without snow around Mexico City.

Ajusco as seen from UNAM. Photo: Ricardo Berlanga. Used by permission

Erdös did not speak Spanish. Fortunately, each of the Mexican students on the trip, in addition to Spanish, spoke another language. Some spoke French, some English, and some German. Erdös was fluent in all of these languages, and he asked each of us what language we preferred. We thought communication was going to be easy. Not quite. Erdös would forget what language each one of us spoke, so he would address someone who spoke French in German, and someone who spoke English in French.

Another interesting aspect of the hike was that Erdös was already in his mid-sixties. Worried about his age, two of our teammates were adamant that we had to climb slowly. What had happened to Witold Hurewicz (who died in Mexico as a consequence of a fall from a pyramid) was already "one too many". Erdös said that he would not have problems climbing up, but, because his ankles were not very sturdy, he would need some help on the way down. Indeed, Erdös did not have any problems climbing up. He went up slowly but steadily, without any help. He did not stop talking and did not

153

stop to rest until he reached the summit. On the other hand, the two teammates who had worried about Erdös, although they exercised regularly, were also heavy smokers. They had to stop a couple of times along the way to catch their breath and then rush to catch up with the group so that others would not notice.

On the top we enjoyed the view. Some of us remembered Gamow's story about a group of Hungarian aristocrats who had lost their way when hiking in the mountains. One of them took a map and after studying it carefully exclaimed. "Now I know where we are! See that big mountain over there? We are right on top of it." (Gamow 1988/1961, p. 3). Knowing where Erdös was from we thought it would be better not to share the story with him. As we started the descent Erdös asked for support. Everybody offered to help. He did not choose any of the strong, athletic young males, but instead chose to lean on a beautiful twenty-year old woman. Go figure.

On the way down, all the young males strode ahead, but Erdös and his support lagged behind, because his ankle was hurting. It did not help that Erdös, instead of hiking shoes, was wearing sandals with socks. A great acclamation and laughter met their return to the base when they finally arrived; the young woman was leading Erdös by the hand. Erdös was a little baffled about the laughter. She did not know how to translate the mocking jokes from her teammates and instead asked Erdös whether he wanted anything. He asked for coffee and we took him to the quesadilla stand. Erdös had two quesadillas and café de olla, the reinvigorating Mexican coffee sweetened with piloncillo (a dark brown sugar rich in molasses).

If someone took pictures along the hike, we did not see them. So we do not have any further proof that we had been with Erdös on a mountain, and thus have Erdös Number 1 for mountain climbing.

References

Erdös Number Project. 2010. Oakland University. Retrieved March 10, 2011 from http://www.oakland.edu/enp/

Gamow, G. (1988). *One, two, three... infinity* (revised ed.). New York: Dover Publications. [Originally published in 1961]

Acknowledgements

Many thanks to Drs. Claudia Gómez Wulschner, Francisco Larrión, Gilberto Flores, and Alejandro Uribe for completing some information and sharing their memories about the trip. Many thanks to Dr. Ricardo Berlanga for taking the picture of Ajusco from the roof of IIMAS and his permission to use it.

Alfinio Flores enjoys teaching mathematics and mathematics education courses at the University of Delaware.

15 A HISTORY OF MATHEMATICS COURSE TO DEVELOP PROSPECTIVE SECONDARY MATHEMATICS TEACHERS' KNOWLEDGE FOR TEACHING[15]

Abstract: Ways are described in which a history of mathematics course can help prospective teachers of mathematics develop knowledge they will need for teaching. The types of knowledge include content knowledge of mathematics, knowledge of mathematics for teaching, and pedagogical knowledge.

Keywords: History of mathematics, pedagogical content knowledge, knowledge of mathematics for teaching, secondary teachers, pedagogical knowledge.

INTRODUCTION

The importance of future mathematics teachers knowing the history of mathematics is highlighted in several professional recommendations for the preparation of secondary mathematics teachers. As articulated in the report *Mathematical Education of Teachers*, "Future high school teachers will be well-served by deeper knowledge of the historical and cultural roots of mathematical ideas and practices" [8, p. 142]. *NCATE/NCTM Program*

---

[15] Huntley, M. A. and Flores, A. (2010). A history of mathematics course to develop prospective secondary mathematics teachers' knowledge for teaching. *PRIMUS*, *20*(7), 603-616. Reprinted by permission of Taylor & Francis (http://www.tandfonline.com).

Standards [27] state that future secondary mathematics teachers should demonstrate knowledge of the historical development of the different mathematical topics, including contributions from diverse cultures. NCTM's professional statements on teachers' knowledge of mathematical content recommend a broad and deep knowledge of mathematical content, processes, and contexts, including "the cultural contexts for mathematics, including the contributions of different cultures toward the development of mathematics" [24, p. 119].

The incorporation of historical ideas into the mathematical preparation of secondary teachers generally takes on one of two forms. One form is to infuse historical ideas into the various mathematics content and pedagogy courses that pre-service teachers are required to take. For example, the books by Fraleigh [14] and Simmons [34] include some historical notes about abstract algebra and differential equations, respectively. Another form is to offer a course devoted to the history of mathematics. Regarding the latter, there are several kinds of history of mathematics courses, each with a different purpose. One option is a course that provides a unified view of undergraduate mathematics [36]. In such a course the mathematics is pursued more thoroughly than in general history of mathematics courses, because mathematics is the goal and history a means to approach the subject as a whole. Another type of course is meant to provide students with a panoramic view of the development of mathematical ideas in approximate chronological order [4, 10, 19, 21]. Other courses may emphasize that contributors to mathematics include women [6, 9], and people from many non-European cultures [18]. Toeplitz [37] outlines a course based on the genetic method, where students witness the origins of problems, concepts, and facts. He selects and utilizes those ideas from mathematical history which came to prove their value. Other courses offer students the opportunity to read the original works of several mathematicians [11], or they may focus on specific aspects of the development of mathematics, such as the history of mathematical technologies [31].

The purpose of this article is to describe a different approach to a history of mathematics course, in which the primary goal is to prepare undergraduate secondary mathematics education majors to become better mathematics teachers. That is, the course has been designed to enhance students' knowledge of mathematical content and ways of thinking about this content vis-à-vis their future teaching. To set the context, the paper begins with a brief overview of the course, including course mechanics and expectations for students. This is followed by further elaboration of how the course prepares students to become better mathematics teachers, which is illustrated with specific examples from the course.

COURSE OVERVIEW

Although any student at the University may enroll in the history of mathematics course, it is a required course for secondary mathematics education majors. Most students take the course in their junior year, which is near the end of their coursework in mathematics and prior to their taking a course in methods of teaching mathematics. Upon completion of their academic program, the secondary mathematics education majors earn a bachelor's degree in mathematics, and most apply for grades 7-12 mathematics teaching certification.

During the first week of class meetings students work with their peers to begin three semester-long in-class projects. One project involves their generating a list of reasons to incorporate the history of mathematics in teaching (e.g., teachers can use historical ideas to motivate and provide deeper understanding of mathematical content). The second project involves their generating a list of ways a teacher can use history in the mathematics classroom (e.g., teachers can demonstrate historical methods using physical objects). The third project involves their generating a timeline for the history of mathematics. So that students can assume ownership of their ideas, volunteers are solicited to post their resulting work on the Internet on a shared course website. Throughout the semester students are expected to revisit their lists of ideas and their timeline, and they are encouraged to contribute new ideas as their thinking develops. At the end of the semester students are encouraged to print a copy of the timeline to post in their classroom to use as a resource in their future teaching.

Students enrolled in the history of mathematics course are required to purchase two types of materials. One material is a CD containing historical modules (lessons) for the teaching and learning of mathematics [20]. The other required material is a textbook [2], which has teachers as one of its intended audiences. The main part of the textbook consists of a collection of 25 short historical sketches. Each sketch focuses on a particular mathematical topic (e.g., π, non-Euclidean geometries), and illustrates the origins of the topic, including key people associated with its development. During the first week of class each student is assigned one of the 25 historical sketches from the textbook. The mathematical topic in each individual's chapter defines the focus of that student's semester-long study into the history of mathematics.

During the course of the semester students examine their assigned mathematical topic from three perspectives. First, students read their assigned sketch from Berlinghoff and Gouvêa [2] and examine other resources relevant to their assigned mathematical topic to understand how their topic was approached in the past, including the historical and cultural influences on the development of the ideas. Students write a 3-4 page paper summarizing their findings and make a 20-minute in-class presentation to share with their peers what they have learned. Second, students explore the life and contributions of a person (usually a mathematician, scientist, or inventor) who has made significant contributions to their assigned mathematical topic and write a 3-4 page biographical sketch of that person. Students are instructed to include in their biographical sketches the historical and cultural influences on the persons with respect to the mathematical ideas they pioneered, and students are required to draw from the *Dictionary of Scientific Biography* [15, 22]. Third, each student uses an historical module from Katz and Michalowicz [20], along with secondary school mathematics textbooks and other curriculum materials, to investigate how her/his mathematical topic relates to current high-school mathematics. Students prepare a 45-minute mathematics lesson on an aspect of their topic, which they present to their classmates. Students are encouraged to include in their presentation aspects of their findings from the prior assignments (i.e., the historical development of the topic and their biographical sketch). In addition to their presentation, students also write a 3-4 page paper summarizing their lesson. In this paper they include reflections on how the lesson played out and explain how knowing the history of their mathematical topic influenced their understanding of the content.

By investigating their assigned mathematical topic from several perspectives, and sharing their findings orally and in writing, over the course of the semester all students are expected to become the "local experts" on their particular topic. Students are graded only on their written work and class participation, not their in-class projects or presentations. Their papers about their topic from a historical perspective, from the perspective of today's high school curriculum, and the corresponding biographical sketch each count for 20% of their grade in the course. Their final paper which integrates these three parts counts for 30% of their grade.

WAYS THE COURSE PREPARES STUDENTS TO BECOME MATHEMATICS TEACHERS

The history of mathematics course was designed to prepare students to become better mathematics teachers. Teachers need several different types of knowledge: knowledge of content, knowledge of pedagogy that is specific to their content area (which Shulman [33] calls *pedagogical content knowledge*), and knowledge of pedagogy. These categories are used in the course to enhance prospective secondary mathematics teachers' knowledge.

Enhancing prospective teachers' knowledge of mathematics

One of the primary goals of the history of mathematics course is to help future teachers enhance their own understanding of mathematical ideas. Two aspects of mathematical knowledge that are important for teachers of mathematics have received attention in recent years. One is developing a profound understanding of the mathematics they are teaching [23]. The other is the knowledge of mathematics for teaching [1]. The history of mathematics course seeks to contribute to these two aspects of mathematical knowledge.

By investigating historical perspectives of mathematical concepts they have already learned, students are presented with alternative approaches to concepts and methods. Students have the opportunity to unpack the content, such as the meaning of π and division of fractions. For example, many students remember from their earlier mathematical learning being asked to memorize the value of the constant π to two decimal places and then practice using it in the formula for the perimeter of a circle, without any explanation of how the formula was obtained or why it works. Through their investigations into the history of π, students in the history of mathematics course gained an understanding of the origins of this concept, namely that ancient Babylonians and Egyptians needed to measure circular objects. Students learned and practiced some ways to compute approximations to its value, such as the rope and stake method and Archimedes' method of exhaustion. Students also gained understanding of why Archimedes used the perimeter of polygons to approximate the circumference, rather than using the area of polygons to approximate the area of a circle. To reach the same degree of approximation, twice as many sides are needed when using the area than when using the perimeter [12]. Students also learned why this same constant appears on both the formula for the perimeter of a circle and for its area. One student remarked, "After researching pi, I have learned that it is not just some mathematical symbol that I learned in high school. It is more than that."

The student who was assigned the topic on quadratic equations also shifted her focus to a more conceptual understanding and meaning of the content, rather than just algorithms. In addition, this student gained an appreciation for connections among mathematical ideas. This student reported that prior to taking the course on the history of mathematics she knew that solutions to equations of the form $y = ax^2 + bx + c$ can be found by using the quadratic formula, $x = \dfrac{-b \pm \sqrt{b^2 - 4ac}}{2a}$. In investigating how quadratic equations were solved in the past, she learned about Al-Khwarizmi's solving quadratic equations of the form $x^2 + bx = c$ by using the rule (written in modern notation) $x = \sqrt{\left(\dfrac{b}{2}\right)^2 + c} - \dfrac{b}{2}$, and she also learned

Al-Khwarizmi's geometrical justification for this formula, which involves completing the square [2]. In her written report this student said:

> Something else that I discovered when doing this research is the connection between algebra and geometry, and how the two intertwine. Until doing this, I had never realized the strong connection between the two, and that this is something that could really help students in seeing mathematics as a big picture, instead of in parts, as I did.

During her class presentation this student made this comment orally, which the instructor used as an opportunity to introduce the notion of integrated versus single-subject curricular approaches to high-school mathematics.

Enhancing prospective teachers' pedagogical content knowledge

Shulman [33] defines "pedagogical content knowledge" as the ways of representing and formulating the subject that makes it comprehensible to others. This includes knowledge of the most regularly taught topics in one's subject area, the most useful forms of representations of those ideas, and the most powerful analogies, illustrations, examples, explanations, and demonstrations. Several aspects of the history of mathematics course enhance students' pedagogical content knowledge.

The breadth of content that is in the textbook [2] and that is discussed in class meetings allows students to review topics in the high school curriculum they have not seen in several years. Moreover, the course assignment to prepare and deliver a 45-minute lesson is intended to promote students' thinking about the secondary school mathematics curriculum from the perspective of a teacher. To complete this assignment, students must rethink which topics are appropriate for high-school students and how to present these ideas. To facilitate their development as future mathematics teachers, one required resource for their papers is the NCTM

Standards [29]. Some students use their writing assignment as an opportunity to reflect more deeply on how they perceived their lesson played out, thus developing their self-assessment skills to be reflective practitioners.

Alternative approaches to teaching

Knowing the history of mathematics can provide future teachers with alternative approaches to the teaching of a topic, and thus offer opportunities to deal with mathematical concepts from different points of view. For example, early Chinese mathematicians, in *Nine Chapters on the Mathematical Art* [32], use a method to divide fractions by first finding common denominators, and then simply dividing the corresponding numerators, so that, for instance, $\frac{2}{3} \div \frac{3}{4} = \frac{8}{12} \div \frac{9}{12} = \frac{8}{9}$. Thinking about why this algorithm works, and how it is related to common algorithms taught in school, can help students develop a better understanding of division of fractions. Future teachers can also study different approaches to solve word problems used in different ages, such as the method of false position [25, 26], or by inverting the operations involved [3, 13]. Again, studying these methods and comparing to algebraic methods used today can provide opportunities to develop better understanding for their students. A popular way for future teachers to incorporate historical ideas into their teaching of mathematics is through the use of games. A more detailed discussion and examples are provided elsewhere [17].

One aspect of pedagogical content knowledge involves understanding ways to motivate students to learn mathematics. The history of mathematics course served as a catalyst for a student studying the development of the ideas of sine and cosine to think about how to introduce this concept to students in a way that piques their interest.

> I believe the greatest challenge as a teacher is motivating students to want to learn and excel in mathematics. Many students are disinterested because they see no relevance to their life, therefore making it unimportant, which is the cause of lack of motivation …. By understanding the history of sine and cosine, students may be able to relate to its significance through astronomy, or the navigation and exploration …or they may see the prevalence of these functions in many fields today, such as, surveying map-making, military artillery calculation and analysis of motion. My new found appreciation for trigonometry, specifically the functions of sine and cosine, will help me be the motivator for my future students.

Some students in high school can be motivated to learn the content by learning about the people who have contributed to the development of mathematical ideas [30], yet oftentimes they are not exposed to the humanistic aspect of the discipline. They have not considered that mathematical ideas are generated by specific people from the past and the present, that mathematics is still being created, and that it is an integral part of human activity. By writing a biography of a person who has made contributions to the development of their assigned mathematical topic, students in the history of mathematics course have the opportunity to examine the human aspect of mathematics and to put a human face behind the ideas.

Depending on the person chosen and the information available, students write a sketch in the spirit of an intellectual biography, focusing on the person's contributions to the development of the mathematical ideas, or they write a humanistic biography, providing detailed information about the person's life, both within and outside of mathematics. Often the choice depends not only on the inclination of the student, but also on the information available about the person. The information on the lives of mathematicians and scientists varies widely. For example, very little is known about the life of Euclid [5]. In that case students would write mainly about the contributions of Euclid to mathematics. In the case of Archimedes, famous anecdotes are told about him, some legendary, some factual, such as his finding a solution while taking a bath to determine whether a wreath had indeed the amount of gold claimed, or his contributions to the defense of his hometown against the Romans [7, 35]. Several of his mathematical works have been preserved; however, we know little about his family and personal development. A student focusing on Archimedes would probably also write mainly about his mathematical contributions while including only a little on the humanistic side. In contrast, in the case of Cardano [16] we have his own autobiography as well as other detailed accounts of his life, his family, his activities outside mathematics (in medicine, astrology, and gambling), and his times. A student focusing on Cardano would have the opportunity to write either a humanistic biographical sketch or one focusing on his mathematical contributions, or both.

At the end of the course, a student who had researched the history of mathematical notation wrote about his strong belief that secondary mathematics teachers have an obligation to include historical perspectives in their future teaching. This student said, "To use these symbols daily without ever learning their origins seems almost disrespectful to the many people who made such outstanding contributions to the way we see mathematics." Indeed, during the course many students gained new appreciation for mathematics and the sometimes long and arduous process surrounding its development. One student traced the development of how whole numbers are written, including the tally system, the Ancient Egyptians' use of hieroglyphic symbols, the Mesopotamians' use of wedge-shaped symbols, the Mayans' use of dots and dashes, the Roman numeral system, and finally, the Hindu-Arabic system with which we are familiar. This student said, "It is amazing to see how much time and effort it took for people from all different places to come up with a system that works well and that everyone can use."

As another example of a student gaining newfound appreciation for mathematics, the student who was assigned the topic of cubic equations remarked that before this course he never really thought about why in high school he was never shown the closed-form solution to a cubic equation. During his class presentation this student traced historical approaches to solving a cubic, and in particular, he explained why Cardano's formula for solving the cubic $x^3 + px + q = 0$ was

$$x = \sqrt[3]{-\frac{q}{2} + \sqrt{\frac{q^2}{4} + \frac{p^3}{27}}} + \sqrt[3]{-\frac{q}{2} - \sqrt{\frac{q^2}{4} + \frac{p^3}{27}}} .$$ After showing how to apply this

formula to a couple of examples, the complicated arithmetic and algebraic calculations that resulted prompted him to say that he now understood why this was not a part of the high-school mathematics curriculum.

Many students intersperse throughout their presentations their own personal observations and reflections about teaching. For example: "When I was observing in my field placement last year I noticed high school students had difficulty with ..."; "I was taught by rote, and I know this isn't a good way to learn, so I included in my lesson a hands-on activity"; and "I think history of math can enliven the classroom, create interest." Such comments provide evidence that the pre-service teachers are thinking about ways to present the content to facilitate understanding, and thus using the history of mathematics course to develop their pedagogical content knowledge.

Enhancing prospective teachers' pedagogy

Throughout the semester most class meetings involve students giving presentations. Each student makes a 20-minute presentation outlining the historical development of their assigned mathematical topic. The other students are required to read in advance the corresponding chapter in the textbook. They ask questions during the presentation and participate in the discussion afterwards. Later in the semester each student draws from secondary mathematics curriculum materials to prepare and deliver a 45-minute mathematics lesson on an aspect of their topic. The other classmates participate in the lesson as if they were high school students, and benefit from learning teaching ideas and activities from the presenter.

Prior to taking this course most students have had very few opportunities to talk in front of an audience. The instructor emphasizes to students that the primary goals of the presentations are to provide them an opportunity to present ideas to their classmates, to help overcome fears, and to develop their poise speaking in front of audiences. Students are not graded on the content or the style of their presentation. Before any students make their presentations, the instructor models one presentation using a topic from the textbook that has not been assigned. Some students ask for the instructor's advice on whether to present their information in didactic fashion or in a more interactive format (which they have learned about in some of their education classes). The instructor advises students to use any format with which they are comfortable.

Students use different strategies in presenting their ideas to their classmates. Most prepare slide presentations using PowerPoint, and occasional glitches in using computers during class provide opportunities for students to learn about the danger of relying on technology when teaching. Some students use the blackboard for their presentations, and some do so with great difficulty. For instance, one student who chose this method for presenting his ideas found it problematic to coordinate his speech with his writing; that is, oftentimes what he wrote was not what he intended to write, and did not match the words he spoke.

It is not surprising that most students' presentations are fairly didactic, given that this course is taught before the methods course, and that most of the experience students have learning mathematics are with this mode of presentation. Students take the presentations very seriously and prepare thoroughly for them. In the course offered along these lines in the Spring of 2008, fellow students were very kind to the presenters. They were gentle with the types of questions they asked their peers, they helped them recover from mistakes, and they seemed to go with the flow while assuming the role of high school students. They even applauded at the end of each presentation.

A short question-and-answer session follows each presentation. As time allows, for instance, when a student's presentation runs short, the instructor provides feedback (praise and areas in need of improvement) about the presentations thus far. At the beginning of the semester this is done in a general way and comments are quite gentle (e.g., I've noticed across several presentations that people are speaking too fast; you need to move away from the board after writing to allow everyone to see what is written). Later in the semester the feedback is more targeted. Students receive positive feedback, for example, in situations where the lesson begins with a discussion of why the topic was chosen, or if the student made explicit connections to others' presentations/topics. Neutral remarks are in the spirit of helping students think about how knowing the history of their topic will inform or improve their future teaching. Remarks given to students to help improve allow the instructor to emphasize some basic teaching principles, such as projecting one's voice so that all can hear, using professional language rather than language that is too colloquial, ways to respond to questions that the presenter does not know the answer, being aware of disturbing other classes nearby, and the purposes of and ways of providing lesson closure. The instructor also addresses issues related to the content, such as focusing on understanding rather than rules and procedures to be memorized, and having learning goals that drive the lesson rather than just presenting "cool activities." Presenters are encouraged to do ongoing assessment by walking around when assigning individual or small group work to get a sense of students' difficulties and successes. Also discussed is the importance of eliciting students' thinking and knowing their audience.

After each student's second presentation, the instructor confers after class in one-on-one fashion with the presenting student. The instructor begins by asking the student his/her perception of how the lesson played out, asking what went well, what did not go so well, and what the student would change if the lesson was taught again. The instructor then shares his/her perceptions of how the lesson played out. This debriefing exercise is the same process students will encounter during student teaching when being observed by their supervisor.

The instructor emphasizes the idea of incremental change both for the written papers and the class presentations. In other words, consistent with best practices in using assessment to enhance learning [28], students are encouraged to reflect on the feedback provided on their assignments (written and oral), and to focus their efforts on improving *only* one or two specific aspects in future assignments. Students are encouraged to make small changes in their writing and presentation styles, rather than making extensive changes all at once. For instance, at the start of students' second presentations the instructor asks them to identify the one thing they want to improve upon with respect to presentation skills. Most students choose language, for example, slowing down their discourse or using more professional language. At the end of their lessons the presenting students, as well as their peers, reflect on the extent to which the presenters were successful in achieving their goal.

FINAL REMARKS

It is common for teacher preparation programs for secondary mathematics teachers to require students to take a history of mathematics course. In this paper we have outlined an approach to a history of mathematics course that is intended to help students become better teachers. Class assignments and class meeting times are deliberately planned to emphasize aspects in the history of mathematics that will help develop prospective teachers' knowledge of mathematics, pedagogical content knowledge, and knowledge of pedagogy. Feedback from students suggests that they enjoy the class and find it relevant to their future role as teachers. An unsolicited remark near the end of the semester from two students illustrates this point: "We just wanted to thank you again for making class both interesting, fun and geared toward what we want to do in the future."

At the end of the semester some students reflected on ways in which they perceived the course as positively impacting their development as teachers of mathematics. In her final paper, one student said:

> After this semester, I have gained a new respect for mathematicians in the ancient times, and have also seen quadratic equations in a brand new light. I am hoping that when I become a teacher, I get the chance to teach quadratic equations as one of my lessons .... I have also realized that using the history of a topic can allow the students to appreciate where it came from, and can also be used as a tool for learning. The historical figures discussed are very intelligent, and bringing the students back into the time of discovery solving problems those mathematicians did puts them on the same level, and can give them confidence in their own abilities. Adding history into a lesson puts a twist on things, and steers away from redundancy of learning processes. It also gives students who may not be as strong in mathematics as others, and students who excel in history, motivation and determination.

These ideas about the benefits of the course for prospective secondary mathematics teachers were echoed by another student:

> This course has been an excellent tool for a future mathematics

teacher. I found that the course instilled quite a bit of confidence for me. Over the last few months I was able to accumulate a tremendous amount of knowledge about Euclidean geometry. I was able to appreciate its place in the past as well as its spot in the present [mathematics] curriculum. I feel that I have become an expert in this area, and I actually look forward to teaching it one day. Confidently and effectively I can now incorporate a historical perspective into my high school lessons. Preparing my own lesson and delivering it in front of my peers felt great. It was an astounding learning experience that I took a lot away from. Teaching my peers builds the confidence that is vital for delivering an effective lesson in an actual high school class. I felt that the teaching experience was excellent, and I am glad that I had an opportunity to do this before my senior year. There is definitely quite a bit that I have taken away from this course that will be utilized in my student teaching placement.

REFERENCES

1. Ball, D. L., and H. Bass. 2000. Interweaving content and pedagogy in teaching and learning to teach: Knowing and using mathematics. In J. Boaler (Ed.), *Multiple perspectives on mathematics teaching and learning* (pp. 83-104). Westport, CT: Ablex Publishing.

2. Berlinghoff, W. P., and F. Q. Gouvêa. 2004. *Math through the ages: A gentle history for teachers and others* [Expanded edition]. Washington, DC: Mathematical Association of America.

3. Bháscara. 1971. "Arithmetic (Lílávatí)." In H. T. Colebrooke (Ed.), *Algebra, with Arithmetic and Mensuration from the Sanscrit of Brahmegupta and Bháscara* (pp. 1-127). Ann Arbor, MI: University Microfilms.

4. Boyer, C. B. 1991. *A history of mathematics* (Revised by U. C. Merzbach). New York: Wiley.

5. Bulmer-Thomas, I. 1971. Euclid. In C. C. Gillispie (Ed.), *Dictionary of scientific biography* (Vol. 4, pp. 414-437). New York: Charles Scribner's Sons.

6. Case, B. A., and A. M. Leggett (Eds.). 2005. *Complexities: Women in mathematics*. Princeton, NJ: Princeton University Press.

7. Clagett, M. 1970. Archimedes. In C. C. Gillispie (Ed.), *Dictionary of scientific biography* (Vol. 1, pp. 213-231). New York: Charles Scribner's Sons.

8. Conference Board of the Mathematical Sciences. 2001. *The mathematical education of teachers*. Providence, RI: American Mathematical Society.

9. Cooney, M. P. (Ed.). 1996. *Celebrating women in mathematics and science*. Reston, VA: National Council of Teachers of Mathematics.

10. Eves, H. 1990. *An introduction to the history of mathematics* (6th edition). New York: Brooks Cole.

11. Fauvel, J. and J. Gray. 1987. *The history of mathematics: A reader*. Hong Kong: Macmillan Press.

12. Flores, A. 2002. If pi were equal to 3.... *Ohio Journal of School Mathematics, 46*, 41-44.

13. Flores, A. 2007. Algebra for a princess. *The Centroid, 33*(1), 5- 8.

14. Fraleigh, J. B. 2003. *A first course in abstract algebra*. (7th ed., historical notes by Victor Katz). Boston: Addison-Wesley.

15. Gillispie, C. C. (Ed.) 1970-1980. *Dictionary of scientific biography*, 16 vols., New York: Scribners' Sons.

16. Gliozzi, M. 1971. Cardano, Girolamo. In C. C. Gillispie (Ed.), *Dictionary of scientific biography* (Vol. 3, pp. 64-67). New York: Charles Scribner's Sons.

17. Huntley, M. A. and A. Flores. In progress. Student-made games to learn history of mathematics.

18. Joseph, G. G. 1992. *The crest of the peacock: Non-European roots of mathematics*. London: Penguin Books.

19. Katz, V. J. 1998. *A history of mathematics: An introduction* (2nd edition). Boston: Addison-Wesley.

20. Katz, V. J. and K. D. Michalowicz. (Eds.) 2005. *Historical modules for the teaching and learning of mathematics [CD-ROM]*. Washington, DC: Mathematical Association of America.

21. Kline, M. 1972. *Mathematical thought from ancient to modern times*. New York: Oxford University Press.

22. Koertge, N. (Ed.) 2008. *New Dictionary of Scientific Biography*. Detroit, MI: Charles   Scribner's Sons/Thomson Gale.

23. Ma, L. 1999. *Knowing and teaching elementary mathematics: Teachers' understanding of fundamental mathematics in China and the United States*. Mahwah, NJ: Erlbaum.

24. Martin, T. S. (Ed.). 2007. *Mathematics teaching today: Improving practice, improving student learning*. Reston, VA: National Council of Teachers of Mathematics.

25. Meavilla Seguí, V. and A. Flores, A. 2006. The rule of false position and geometric problems. *Convergence, 1*. On line journal http://mathdl.maa.org/mathDL/1/ . Accessed January 28, 2009.

26. Meavilla, V. and A. Flores. 2007. History of mathematics and problem solving: A teaching suggestion. *International Journal of Mathematical Education in Science and Technology, 38*(2), 253-259.

27. NCATE/NCTM Program Standards (2003). Programs for initial preparation   of   mathematics   teachers. http://www.nctm.org/uploadedFiles/Math_Standards/NCTMSECO NStandards.pdf . Accessed July 7, 2008.

28. National Council of Teachers of Mathematics. 1995. *Assessment standards for school mathematics*. Reston, VA: National Council of Teachers of Mathematics.

29. National Council of Teachers of Mathematics. 2000. *Principles and standards for school mathematics*. Reston, VA: National Council of Teachers of Mathematics.

30. Reimer, W. and L. Reimer. 1992. *Historical connections in mathematics: Resources for using history of mathematics in the classroom*. Fresno, CA: AIMS Educational Foundation.

31. Shell-Gellasch, A. (Ed.). 2007. *Hands on history: A resource for teaching mathematics*. Washington, DC: Mathematical Association of America.

32. Shen, K., J. N. Crossley, and A. W.-C. 1999. *The nine chapters on the mathematical art: Companion and commentary*. New York: Oxford University Press.

33. Shulman, L. 1986. Those who understand: Knowledge growth in teaching. *Educational Researcher, 15*(2), 4–14.

34. Simmons, G. F. 1991. *Differential equations with applications and historical notes* (2nd ed.). New York: McGraw-Hill.

35. Stein, S. K. 1999. *Archimedes: What did he do besides cry eureka?* Washington, DC: Mathematical Association of America.

36. Stillwell, J. 2002. *Mathematics and its history* (2nd ed.). New York: Springer-Verlag.

37. Toeplitz, O. 2007. *The calculus: A genetic approach* (Reprint edition). Chicago: University of Chicago Press.

ACKNOWLEDGEMENTS

The authors express appreciation to Dr. Kathy Clark, Florida State University, for helping to conceptualize the development of this course.

16 USING STUDENT-MADE GAMES TO LEARN MATHEMATICS[16]

Abstract: First year university students design and play their own games, including board, computer, and other kinds of games, to learn mathematical concepts and practice procedures for their pre-calculus and calculus courses.

Keywords: Mathematics games, algebra, trigonometry, geometry, calculus, meaningful practice, motivation, alternative assessment.

INTRODUCTION

In this article we describe how first year college students can design and use their own games to learn mathematics. Games in the mathematics class can be used to practice computational skills, develop better understanding, and foster positive attitudes towards mathematics. Games have long been used effectively to learn mathematics at the high school level [6] and at other grade levels [1, 7, 9]. Many of the same principles that make games effective in schools are also valid with college students. Games can be geared toward all ability levels. Some of the games that are successful with high school students can be easily adapted for college students by incorporating appropriate mathematical content [2, 3]. Students can also use their creativity to adapt or invent new games to learn mathematics. Having students create games and analyze what a good game entails has the potential to induce self-reflection on their learning process.

---

[16] Gallegos, I. and Flores, A. (2010). Using student-made games to learn mathematics. *PRIMUS*, *20*(5), 405-417. Reprinted by permission of Taylor & Francis (http://www.tandfonline.com)

GAMES DESIGNED BY STUDENTS

This article describes experiences college students had constructing and using games to learn mathematics. The students were engineering students in their first year at the University of San Luis Potosí (UASLP) in Mexico. The students worked in small cooperative groups to design their own games to learn mathematics in their pre-calculus and calculus courses [4, 5]. The purpose of each game is to learn one of the topics of the course. Before asking students to design their own games, the instructor brings a game to learn mathematics to class and the whole group participates. After the game, students discuss what they thought about learning mathematics this way. They usually find the experience to be dynamic and voice their opinion that this kind of activity should be enacted in class to develop and reinforce understanding of the topics in class. They also find that playing the game allows them to get to know other students in the class and support each other. The instructor then asks students to participate in the design of their own games to learn mathematics. Students start to brainstorm about what games they would like to design and they work on those ideas without further assistance from the instructor. Once a group of students agrees to work together the instructor provides weekly feedback on their progress. Students can also ask the instructor before or after any class if they have any questions. Games have to satisfy the following requirements.

1) About 90% of the game content is related to one or more of the topics of the course.

2) About 10% are dynamic strategies to enrich the game, such as trick questions, rewards, and reasonable sanctions and punishments.

3) The game has a title, rules, and instructions on how to play it. It should include all the necessary materials for play. There should be notes and comments about what they learned by designing and playing the game. An evaluation form is required so that users of the game can write comments about what they did or did not like.

4) Its size has to be such that it is portable. It should be made of sturdy materials, and have an attractive appearance.

5) Bibliography or references must be included.

Examples of games designed by students

To design their games, students adapt popular board games or other games such as domino, expand and improve games from previous semesters, or use their own ideas. Students also develop games that are played on computers. Students make up the games for their peers as part of a class assignment, but some of the games are also used in later years with students who take the same course. Some of the games are intrinsically of mathematical nature, like the ones discussed by Schuh [8]. Other games are adapted from non-mathematical games, and the kind of questions students choose to write is what makes them appropriate for a mathematics class. Figure 1 shows a version of the Mexican game of *Lotería* (similar to bingo). One student reads the functions on the cards, and the other students have to match each function with the corresponding derivative on their boards. When students have a match, they place a marker on the board. The first student to completely fill the board is the winner.

In Figure 2, an algebraic domino is illustrated. Rather than matching number of dots on different tiles, students need to match corresponding algebraic expressions, for example conjugate binomials $(a + b)(a - b)$ on one tile with a difference of squares $(a^2 - b^2)$ on another. If the student does not have any domino tile that matches, he or she takes one from the extra tiles provided. Surprise tiles are also included, where students lose a turn or gain an extra turn. In Figure 3, a different algebraic domino is shown. Students have to match equations with their solutions, for example, $\sqrt{2x-10} - \sqrt{x+9} = 2$ on one domino tile with $x = 55$ on another tile.

In a board game (Figure 4) students take turns to work on problems indicated on the cards. Students use a die with six numbers to determine the order in which they will play and to see how many places they advance initially, and a die with three colors, which determines the kind of questions. Students advance on the board as many places as indicated on the card when they simplify correctly an algebraic expression like $\dfrac{x^3 - 6x^2}{x^2 - 12x + 36}$ or $(x^4 - x^2 - 2x - 1) \div (x^2 - x - 1)$. If their answer is incorrect they go back as many places as indicated on the card. If the card shows a crossed-out die the student loses a turn. The first student to reach the end is the winner.

In the game *Fishing problems* (Figure 5), students fish problems to practice how to factor out algebraic expressions like $x(a - 1) + y(a - 1) - a + 1$, or $8 - (x - y)^3$ and solve special products by using basic patterns of factorization. If the player makes a mistake, or is unable to factor the expression in the allotted time the player will have to return the fish, loose a turn, and receive a punishment. If the problem is solved correctly, the player will keep the fish and will be allowed to fish again in their next turn. The first player to get three fish will be the winner.

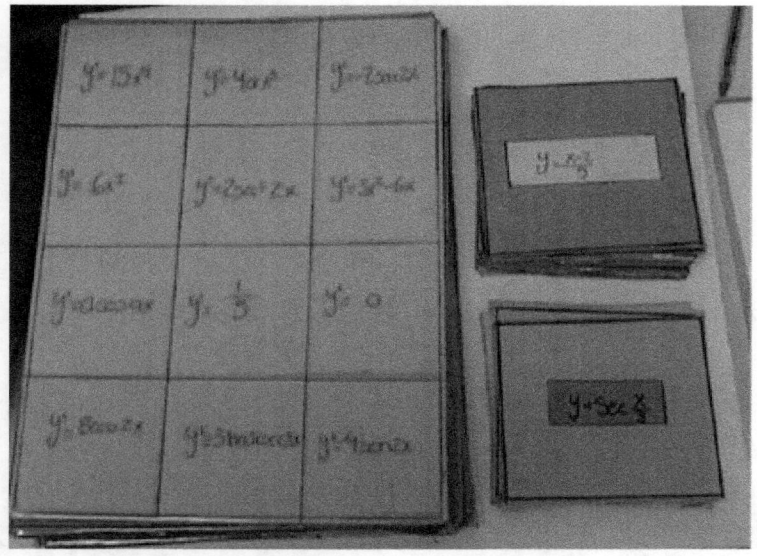

Figure 1. *Lotería* for derivatives. Ability to differentiate.

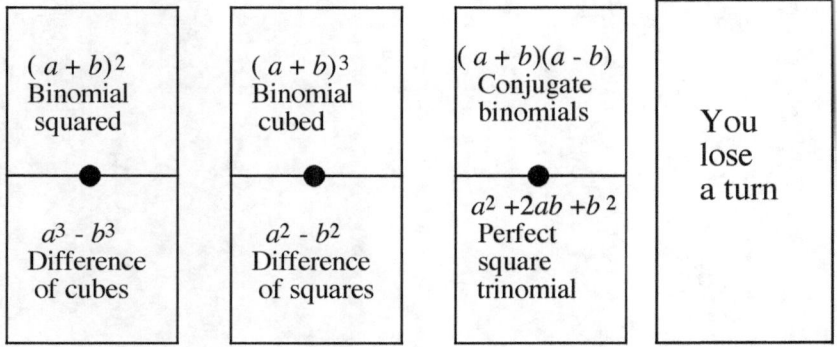

Figure 2. Domino of algebraic expressions

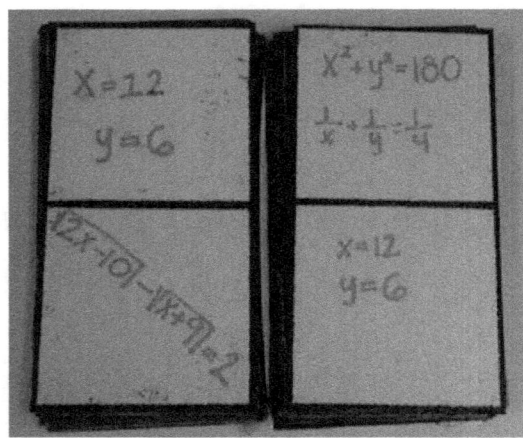

Figure 3. Domino of algebraic equations

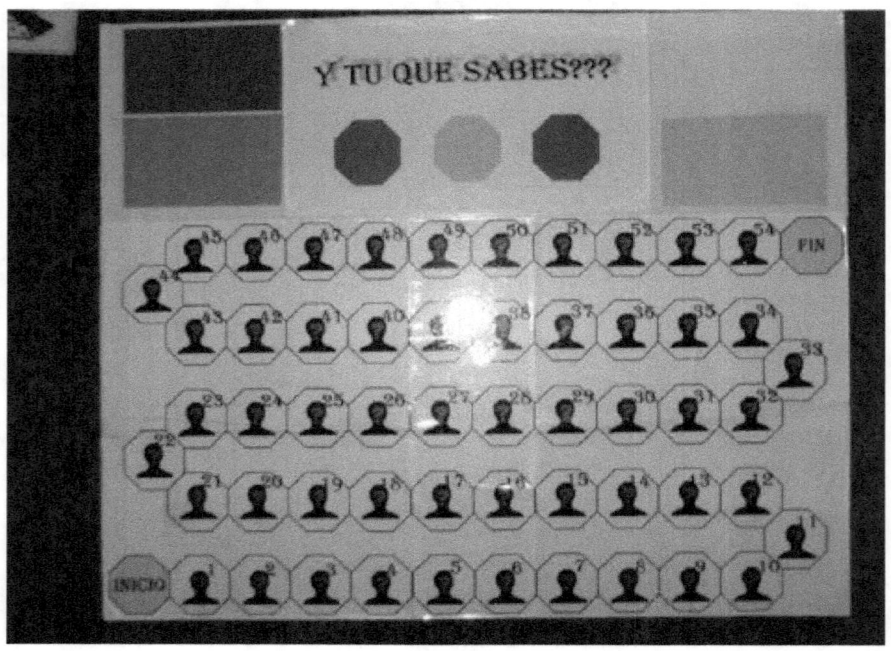

Figure 4. Board for *What do you know?*

Figure 5. *Fishing problems* set

The games described so far focus on practice of skills. The next two games are more geared towards development of conceptual understanding. In a game on geometric proofs designed by students that is played on the computer, players have to prove that triangle ABC is isosceles given that ADEB are collinear, CD is congruent to CE, and AD is congruent to EB (Figure 6). Students have the opportunity to choose the right justification for each step in the proof from two options provided by the computer. The computer offers a sound justification and a faulty one. Players will proceed to the next step if their answer is correct. If their choice is incorrect they are told to study more and go back to the beginning of the game.

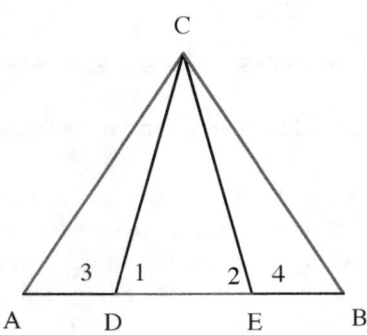

Figure 6. A game on proofs

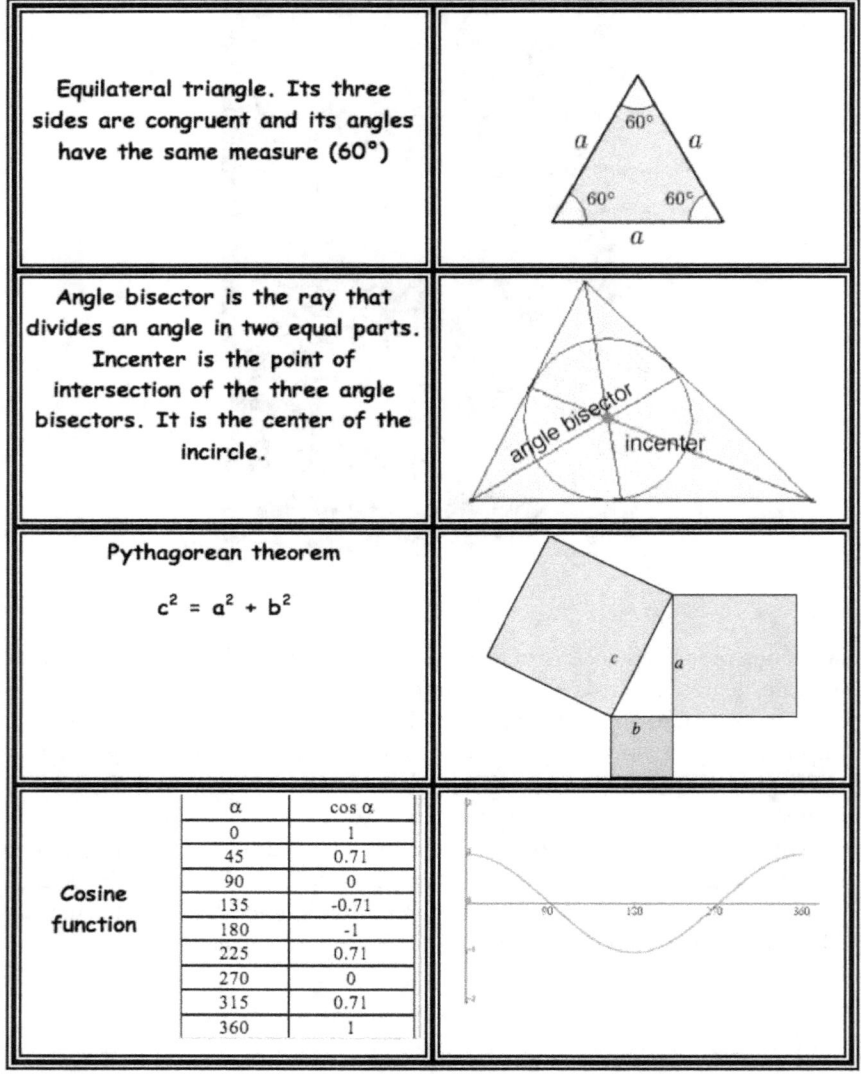

Figure 7. Matching cards for concentration game

The following is an example where the game creates a situation that contains mathematics in iconic form, invites reflection, and encourages mathematical communication and connections. Figure 7 includes some of the cards for a memory game. Students have to form pairs of matching cards, where one card has the definition or name of a concept or the statement of a theorem, and the other has a picture of that concept or theorem, or where one card shows the name and table for a function and the other shows the graph. Cards are placed face down and a student turns two cards over. If the cards match, the student keeps the cards and turns two other cards over. If the cards do not match, the cards are turned face down again and another student repeats the process. The student with the most pairs wins.

Factors for successful use of games

Like any instructional method or material, the success of mathematical games depends on how they are used. The games we have presented share several factors of successful games as described by Johnson [6]. The games involve mathematical abilities and concepts that are important and closely related to the content of the course. All students participate actively. An element of chance in the games gives everybody the opportunity to win and not just the more able students. Games are used during the regular class session when the ideas or skills are being taught. The games are relatively short so that students do not lose interest. The games are planned carefully and conducted in an organized way. The informality and excitement does not interfere with the purpose of the game. Students who play the game accept the responsibility of learning mathematics from the game. The instructor also evaluates to what extent the game is a success in promoting the desired learning.

Preventing and addressing problems

Usually there are very few problems among students working together in the small groups to develop the games, because students choose with whom they want to work, the materials they will use (according to their resources), and where they will develop the activity. The weekly monitoring by the instructor also helps address problems in a timely manner. Asking students about how they feel, both cognitively and affectively, helps them to bring out comments and address any possible misunderstandings. Teams can remain together all semester long or they can change at each of the four exams per semester. The instructor suggests changing roles in the team such as leader, secretary, support person, and so on, so that all students can participate in various ways. If any problem arises in one of the groups, the instructor talks to the small group to find out whether there are problems with one or two of the members. They have the choice to work things out and remain in the group or to ask other groups to accept them. If the whole small group is not working well, the instructor discusses with them what interferes with the work, whether they can change the work they are doing, or if they can split in two groups.

Evaluation of student designed games

Ten percent of the grade for the course is based on the games (four per semester) designed by the students. The games are designed and used before each one of the four exams. For each game students are given formal feedback through an assessment rubric, and an "analysis of performance table". We will briefly describe what each one entails and some of its benefits.

Evaluation rubric. One rubric is used for each game as the work progresses and when the work is turned in. Feedback for students is given at each stage of progress and when the work is completed. This sheet is given to the students so that they can record their own progress and so they know how many points each part is worth. For the instructor, it is a guideline to evaluate the contributions of the students. The team of students and the instructor review the rubric each time there is some progress made until the work is turned in. Each aspect of the game receives a certain number of points. For example, attention to content and length receives up to three points from a total of ten points, according to the following rubric:

The length is as required and the topics are complete (3 pts.)	The length is half that required and the topics are complete (2 pts.)	The length is one third that required and the topics are not complete (1 pts.)	The length and the content do not address the requirements (0 pts.)

The evaluation rubric is very effective with respect to the information given. Students easily recognize what they need to do to complete an assignment, and also what the corresponding weight for each part is. The evaluation rubric is especially important as formative evaluation while students show their progress. The instructor can clearly communicate what has been covered in the assignment and what is still missing. It gives confidence to the students about their progress and about the grade they will receive.

Table of analysis of performance. One table is used for each game when it is played. The instructor registers her observations about the abilities of the students during the activity. Students receive feedback for each game. Development of skills is recorded on this table. When the game is played the instructor observes the abilities that students may develop as they play the games, for example, ability to communicate. One of the parts of the assignment is a form to register the comments of the students about the game they are playing. The instructor takes note of these comments and in addition writes her own observations on the table of analysis of performance. Once the activity is completed the instructor gives feedback to each student who designed the game. Examples of students comments, instructor observations, and abilities developed are presented in Table 1.

The table of performance analysis facilitates communication between the instructor and the students about the development of their abilities. When students describe the impact that the instructional material or game has on their knowledge, they record their trajectory, development, and progress, which are complemented by the observations made by the instructor.

Table 1. Students' comments and instructor observations

Student comments	Instructor observations	Ability development
In this task we have learned simple [algebraic] fractions in an easy and fun way. I would include some rules to organize better the actions.	Contributes new strategies, work guidelines, or ways to do things.	Leadership
This game allowed us to understand the important concepts of simple [algebraic] fractions. Invites the participation of all and with that one feels more secure in what is learned.	Does address effectively the challenges he faces because he has the support of his group and knows how to take advantage of combined talents.	Team work
I would like that this game would include the basic structure of factorization in a table where the team can look at it to reaffirm the knowledge and make the exercises more practical.	Poses creative ideas to solve any problem.	Innovation
With this activity several ways of solving can be followed. It is fun and you learn.	Knows how to communicate with his team and persuade about the strategy to be used.	Ability to persuade and communicate

Playing the games in class

In a semester when students design games, two full class sessions are spent playing games in class. Four sessions are used in semesters when students play with games designed by students in previous years. In a session several games are played at the same time in small groups, and when the game is finished students rotate and play a different game. If the game is long they may play only one game in a session. Students name a guide for each game who explains the rules, serves as referee, and keeps the game moving. Later, all students come together and a game is chosen to play as a whole group. The instructor rotates from group to group, checks that the activity is done by each group, provides support if students have questions, and serves as referee if needed. She checks the changes in attitudes that students have and the roles they play. She always takes notes, photographs, and sometimes video to gather evidence.

Students actively and enthusiastically participate in the games in class (Figure 8). They make their best effort to get the right answers to do well in the game. To make the game more fun, students sometimes include in their games small rewards for correct answers such as receiving a piece of candy, or half a point on the next exam, and slightly embarrassing punishments for wrong answers such as telling their latest folly or worst nightmare.

Figure 8. Active participation

Benefits of designing and playing games

The design and use of instructional materials and educational games offers several potential benefits for the students. The instructor gather evidence as she interacts with students and observes them interact with each other. Students feel more motivated and interested in mathematics. They feel better adapted to the new world that is the university, where they have recently arrived and where integration is difficult at the beginning of the first semester or even the second semester. Students need to apply their knowledge to design the games. Students face different situations in life where they can play a successful role in the cognitive realm, develop the courage to face challenges, and be successful with their own resources.

We end this section with some examples of students' comments about the benefits of playing games to learn mathematics.

- It is a practical, simple, and fun way to learn to review the knowledge acquired in class.
- This game helped us, in addition to reaffirming the knowledge acquired in class, to play with some expressions and problems to be able to understand them better.
- The project has served to give importance to work in teams, and above all, reinforce the topics of the class.
- Looking for an improvement of the project made us get to know each other better as classmates, and help us share our knowledge in a different way.

FINAL COMMENTS

There are some cases in which it is not easy to introduce the approach presented here. In some groups, students' apathy, and lack of interest or outside distractions, make it hard to implement. In any case, it is our intent to provide support for students who show an interest in improving themselves. Students' comments as they play the games make participation more dynamic, stress positive aspects of the games, help to correct any mistakes in the directions or exercises of the games, and suggest ways to improve the games from one semester to the next.

REFERENCES

1. Bright, G. W., J. H. Harvey, and M. M. Wheeler. 1985. *Learning and mathematics games*. Reston, VA: National Council of Teachers of Mathematics.

2. Flores, A. 1991. A puzzle of mathematical formulas. *PRIMUS, 1* (4), 397 - 400.

3. Forman, S. and S. Forman. 2008. Mathingo: Reviewing calculus with bingo games. *PRIMUS, 18*(3), 304-308.

4. Gallegos, I. 2006. "Material Didáctico y Juegos Educativos" In Memorias del XIX Congreso Nacional de la Enseñanza de las Matemáticas, ANPM. Chihuahua, Chih.

5. Gallegos, I. 2007. "Evaluación juegos educativos." Paper presented at the 12a Conferencia Interamericana de Educación Matemática, Querétaro, Qro., México (July).

6. Johnson, D. A. and G. R. Rising. 1972. *Guidelines for teaching mathematics* (second ed.). Belmont, CA: Wadsworth Publishing Co.

7. Kohl, H. R. 1974. *Math, writing, & games in the open classroom*. New York: New York Review.

8. Schuh, F. 1968. *The master book of mathematical recreations*. New York: Dover.

9. Smith, S. E. and C. A. Backman, (Eds.). (1990). *Games and puzzles for elementary and middle school mathematics*. Reston, VA: National Council of Teachers of Mathematics.

17 ADVICE FOR NEW STUDENT TEACHERS FROM BEGINNING TEACHERS[17]

At the end of the semester, during the last session of the weekly seminar that accompanies the student teaching experience, we asked our student teachers to think about the one piece of advice they would give to beginning student teachers. We went around the circle and everyone took their turn to express their thinking. They listened carefully to each other. In some cases they complemented or expanded on previous advice. In other cases they took the thinking into a different aspect of teaching. Their poise and thoughtful advice confirmed that one of the most amazing transformations had taken place before our eyes in less than one year. The group of students—enthusiastic about math but somewhat naïve about teaching and learning mathematics—we met the previous fall at the beginning of the methods course had transformed itself into a group of beginning teachers with wonderful experiences and valuable insights about teaching. We thought that what they said would be worthwhile to collect and share, not only because it was indeed very valuable advice for future student teachers, but also because such a list would be good for them to have present at the beginning and during their first year as new classroom teachers. Furthermore, we thought that even for experienced teachers it would be a good list to have in mind. We have both have long experiences teaching, the first author as a high school teacher, and in teacher preparation programs, and the second author as university professor and in professional development of teachers. As we were listening to what each of

[17] Biondi, K. and Flores, A. (2010). Advice for new student teachers from beginning teachers. *Ohio Journal of School Mathematics*, No. 61, 4-6. Reprinted by permission.

the student teachers was emphasizing, we realized that much of what we had found effective in teaching over the years had been encapsulated succinctly and to the point by our students. As we were compiling their comments, we noticed three distinct classifications seemed to emerge: connections with students, classroom procedures, and personal survival skills. These are the suggestions our students offered to new student teachers.

Connections with students

- Get to know your students.
- Talk with your students about their outside activities and interest.
- Attend a variety of student activities outside of the classroom.
- In the beginning while you are still observing classes, target disruptive students and work with them individually.
- Coach, if possible.
- Remember that students are just kids that you are trying to mold into adults.
- When students make negative comments to you, don't take them to heart.
- Don't assume students know too much. When you first begin, you will expect students to know a lot more than they actually do.
- Avoid making negative comments about a student or criticizing a student, especially in the faculty lunch room.

Classroom procedures

- Plan ahead. Work on something every night even if you don't need it for the next day.
- Don't sweat the small stuff. If you make a mistake on the board, don't stress about it.
- Stay consistent (rules, grades, etc.).
- Know the limitations of your responsibilities when it comes to students.
- Stay organized, especially with records such as student absences and students who are missing a test.
- Make copies of handouts prior to when you need them. If you wait until the morning of the day you plan to use them, the copier may not be working.
- Have a backup plan. For example, you may plan to use the overhead projector but the bulb burns out.

- Don't be afraid to deviate from your lesson plans. Students may ask a question or make a comment which is relevant to the topic but which takes your lesson in a different direction than originally planned.

Personal survival skills

- Enjoy your students. Show enthusiasm.
- Be yourself. Don't copy someone else's teaching style.
- Interject humor.
- Laugh at yourself.
- Relax and enjoy student teaching.
- Student teaching is a learning experience. Don't get down on yourself.
- Relax. Things will fall into place.
- Ask yourself, "Would I want my own kid to be in my class?"
- Clear up your schedule. Teaching is a lot of work and very time consuming.

Final remarks

We noted that none of the comments had any direct reference to mathematics. In part this could be due to the fact that secondary mathematics students at the University of Delaware do have strong backgrounds in mathematics. They are quite skilled in high school mathematics and go through a rigorous program of content courses. During the mathematics methods course they realized that much of their knowledge was procedural, and that often they had learned the *how*, but not the *why* of many procedures. However, in that course they developed the ability to figure out on their own the reasons for much of what they had learned and were able to fill any missing gaps in their knowledge of high school mathematics. Because of this, most of our students felt quite confident with respect to their mathematics knowledge at the beginning of their student teaching experience. In contrast, the opportunities to actively interact with secondary students before student teaching at Delaware are not as abundant or systematic as in other programs. Opportunities to develop classroom management skills, as well as getting to know adolescents closely were plenty during student teaching, rather than before. As a result, the advice our students provided focused on the aspects they themselves had to pay the most attention to at the beginning of their own student teaching experience.

Acknowledgements

This article is based on advice given by Josh Bowman, Kevin Cauto, Lauren Durie, Jamie Flynn, Kayla Freeman, Megan Henley, Amy Holly, Emily Hoover, Greg Inman, Larry Isakoff, Shon Jablonksy, Laura Kleinstuber, Marianne Lios, Brian Marley, Angela Pollino, Mark Roche, Lisa Sherry, Julia Somers, Kevin Stark, Tony Webb, Sarah Zilberfein, Carrie Zwaan.

18 AREA FORMULAS WITH HINGED FIGURES[18]

Students often learn the formulas for areas of figures such as rectangle, triangle, parallelogram, trapezoid, and regular polygons without making connections among them. Often, too, students do not realize that the algebraic transformation of one area formula into a different equivalent expression, for example $\frac{bh}{2} = \frac{b}{2}h$ can sometimes be interpreted in geometric terms as well as algebraically. The goal of this chapter and its corresponding web site is to help students make such connections with the use of interactive hinged figures. These hinged figures may be manipulated in different ways. Some are transformed by rotating some of their parts to form other shapes that have the same area. Others are rotated as a whole to form, together with the original, a new figure. By comparing the original figures to the new figures students can establish relationships among different formulas for areas. As they rotate the figures or their parts, students can get a kinesthetic as well as a visual sense of what parts of the original figure correspond to parts of the transformed figure. The guidance of a teacher is critical in helping students interact with the figures in a way that will maximize their understanding.

The interactive figures can be found at http://www.public.asu.edu/~aaafp/yearbook/algebrahinged.html. Readers are encouraged to interact with these figures before or during their reading of this article.

---

[18] Flores, A. (2009). Area formulas with hinged figures. In T. V. Craine (Ed.), *Understanding geometry for a changing world* (pp. 297- 313). Reston, VA: National Council of Teachers of Mathematics. Copyright National Council of Teachers of Mathematics. Used by permission.

We will briefly discuss how this interaction can play out when the emphasis is on the learning of geometrical ideas per se and then discuss how the interactive figures and the formulas for areas can provide a bridge to algebraic thinking and notation. Finally we will look at each of the figures and indicate the questions that may be asked to get students to engage with them.

Several of the arguments of this chapter depend upon the fact that when a segment is rotated 180° about its midpoint, it coincides with itself. This may be demonstrated by letting students experiment with concrete straight objects such as toothpicks or popsicle sticks to represent segments which can be rotated about their midpoints. The interactive figures then provide an additional experiential setting where students can see what properties are preserved when figures or their parts are rotated in other ways. They can see, for example, that a segment rotated 180° around a point not on the segment will give another segment that is congruent and parallel to the first. They can also see relations that emerge from the way the figures are partitioned, for example, that the segment connecting the midpoints of two sides of a triangle is parallel to the third side. For older students who are learning to develop deductive arguments in geometry, the interactive figures can suggest conjectures and provide leads to understanding why they are true. Students can then prove the relations between the original and the transformed figures by using traditional arguments of congruency or by basing their arguments on properties of rotations previously proven or accepted. Students can also see how the formulas for areas can be organized into a deductive system in which some formulas are derived from others.

When the emphasis is on bridges to algebra, students can work through a sequence of geometric figures with the same area and find corresponding algebraic expressions for each figure and its parts. The different parts of the algebraic expression correspond to parts of the figure in each case. Alternatively, teachers can give students a sequence of algebraic expressions that are equivalent and ask students to find geometric interpretations for each term or factor in the different expressions. Area formulas can thus provide a context for meaningful manipulation of algebraic expressions. On one hand, students have the opportunity to interpret geometrically the parts of the different formulas, and see why they are equivalent. On the other hand, students can use algebraic principles such as the associative, commutative, and distributive properties to transform one algebraic expression into another. Whereas often students simply learn to substitute specific numerical values into the formulas, here the purpose is to treat the different elements in a formula as mathematical objects per se. In some cases there are subtleties in the algebraic reasoning required and the teacher may want to make them explicit. Students with a wide range of mathematical backgrounds can learn about areas and geometrical relations by interacting with the figures. Teachers can provide help and guidance according to the mathematical maturity of the students to develop the more subtle algebraic aspects.

There are questions posted on the web site for each of the interactive figures to help students focus their attention on relevant relationships. However, the teacher still needs to play an active role guiding students so they realize the area formulas that are being illustrated in each interactive figure. Appropriate and timely prompts from the teacher support students in seeing the relationships, describing them in their own words, and using mathematical notation.

The basic formula for area on which all the other formulas in this article are based is the area of a rectangle. To make the connection more explicit, instead of using length and width to describe the rectangle, we say that the rectangle has base b and height h, so that the area formula is given by $A = bh$.

Triangle

For the area of triangles, the simplest case is the right triangle. The Interactive Sample Figure in the Instructions shows one example. The hinge or pivot is at the midpoint of the hypotenuse. The triangle is rotated 180° around the hinge (Figure 1). Teachers can help students make explicit the mathematical relations by asking questions to focus their attention on properties and relations of different parts.

- What is the area of the original triangle compared to that of the rotated triangle?
- What shape do the original right triangle and the rotated triangle form together? How can you justify your claim?
- What is the relation between the area of the triangle and the area of the rectangle?
- If the base of the rectangle (its length) is b, and its height (width) is h, what is the formula for the area of the rectangle?
- What should be the formula for the area of the right triangle?

The goal is for students to express the area of the right triangle as A

$$= \frac{1}{2}bh.$$

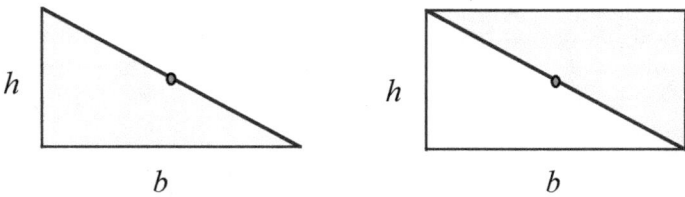

Figure 1. Right triangle and rectangle

The rectangle can also be used to derive the formula for more general triangles. In the next case the altitude of the triangle falls in the triangle's interior (Figure 2). In interactive Figure 1 the hinges are at the midpoints of two sides of the triangle. The segment with measure h is perpendicular to the base of the triangle and forms two right triangles. Teachers can again ask questions as students interact with this figure so that they can make explicit the relations among parts,

- How are the areas of these two triangles together related to the area of the original triangle?
- What figure do the two rotated parts form together with the original triangles? How can you justify your claim?
- How can we justify that the new figure is indeed a rectangle?
- What is the area of the original triangle compared to the area of this rectangle?
- What is the base of this rectangle in terms of the original triangle?
- What is the height of this rectangle compared to the height of the original triangle?
- Express the relation between the areas of the original triangle and the rectangle verbally using your own words.
- If the area of the rectangle is given by bh, write a formula for the area of the triangle.

The goal is for students to express the area of the triangle as $A = \frac{1}{2}bh$. Of course, students can also say that the area of the triangle is equal to the area of the rectangle divided by 2, and write it as $A = \frac{bh}{2}$.

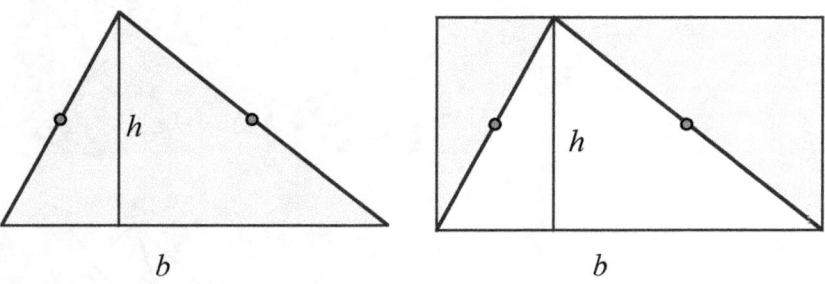

Figure 2. Triangle with inside altitude

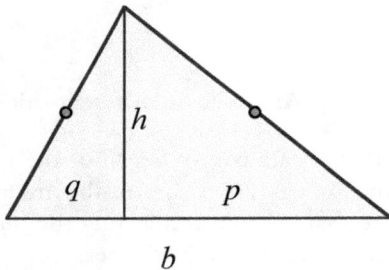

Figure 3. Sum of areas of right triangles.

Students can also describe how the triangle is formed by two right triangles and use letters to label the parts and algebraic notation to express the relations. If an altitude of measure h cuts the base of the triangle in parts q and p (Figure 3), what are the bases of the two right triangles? What is the sum of the bases of the two right triangles compared to the base of the original triangle? $(q + p = b)$ Students can express the area of the triangle as the sum of the areas of the two right triangles, and simplify the algebraic expression using the distributive and commutative properties:

$$A = \frac{1}{2}qh + \frac{1}{2}ph = \frac{1}{2}h(q+p) = \frac{1}{2}hb.$$

Every triangle has three altitudes and the area may be found with any side as base and the corresponding altitude. In interactive figure 1 a triangle is shown to illustrate the above argument. Of course the argument could be made using any of the three altitudes (Figure 4). This may help students overcome the misconception that there is only one altitude, and that it must be associated with a side that appears horizontally.

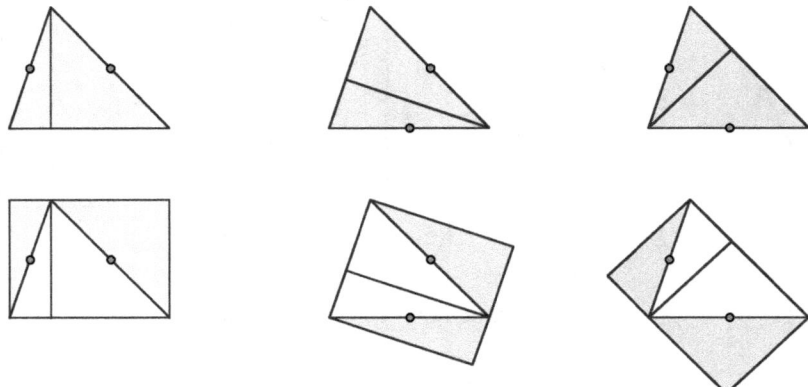

Figure 4. Altitudes on different sides.

A different approach is to rotate parts of each of the two right triangles. In interactive figure 2 the hinges are again at the midpoints of two sides. Students need to realize that the line connecting the midpoints is parallel to the base, perpendicular to the altitude with measure h, and cuts the altitude in half (Figure 5). The teacher may want to ask students to justify each of these facts. Here are some other questions teachers can ask to make the relations explicit as students interact with the figures.

- What kind of shape do the two rotated triangles form together with the section of the original triangle that did not move? How can you justify your claim?
- How does the base of this rectangle compare to the base b of the original triangle?
- How does the height of the rectangle compare to the height h of the original triangle?
- If we write $\frac{h}{2}$ for the height of the rectangle, what would be a formula for the area of the rectangle?

The goal is for students to see that the area of a triangle is equal to the area of a rectangle with the same base and half the height, $A = b \times \frac{h}{2}$.

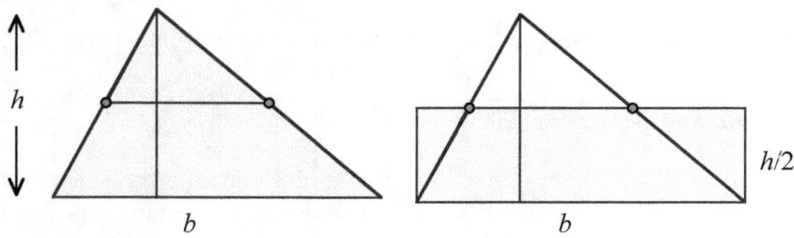

Figure 5. A rectangle with half the height

In interactive figure 3 the hinges are again at the midpoints of the sides, and two small right triangles are formed by segments through the midpoints perpendicular to the base of the triangle. The segments through these points are thus parallel to the altitude and will cut each of the segments of the base of length p and q in half (why?) (Figure 6). Teachers can guide students to express the relation between these half segments and b using algebraic notation, $\frac{p}{2}+\frac{q}{2}=\frac{p+q}{2}=\frac{b}{2}$. The teacher may also have students explain why the new shape is a rectangle. Students can thus see that the area of a triangle is equal to the area of a rectangle with the same height and half the base, $A=\frac{b}{2}\times h$.

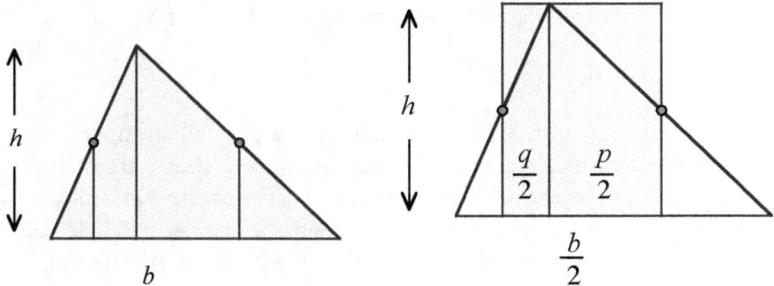

Figure 6. A rectangle with half the base.

We have thus different algebraic expressions for the area of the triangle. $\frac{1}{2}bh=\frac{bh}{2}=b\times\frac{h}{2}=\frac{b}{2}\times h$. The first two express essentially the same geometric fact; each of the last two has a different geometrical interpretation. The teacher may also want to make explicit how the commutative and associative properties are used in transforming one formula into the other, for example, $b\times\left(\frac{1}{2}\times h\right)=\left(b\times\frac{1}{2}\right)\times h$ by the associative property, and $b \times \dfrac{1}{2} = \dfrac{1}{2} \times b$ by the commutative property.

Rectangle and parallelogram

In interactive figure 4 the hinges are at the midpoints of the sides of the parallelogram. When the small triangles are rotated a rectangle is formed (why?). The base of the parallelogram has the same length as the base of the rectangle formed by rotating the two triangles; students see this visually or by reasoning as follows. If c is the base of the small triangle, the base of the rectangle will be $b - c + c = b$ (Figure 7). The area of the rectangle is thus bh and therefore, the area of a parallelogram is given by $A = bh$. This approach works for parallelograms where the projections of the midpoint intersect the bases. An approach that works with all parallelograms, including the "long skinny ones," is to use the connection with the formula of a triangle, as shown in the next section.

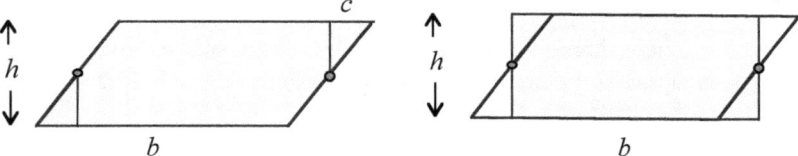

Figure 7. A rectangle with the same base and same height.

Triangle and parallelogram

We can find another route to the formula of the parallelogram by using the formula for the area of a triangle. Interactive figure 5 shows that two copies of an acute triangle make a parallelogram. Teachers can ask questions so that students explicitly understand that indeed the rotated triangle together with the original form a parallelogram, that the base of the triangle is the same as the base of the parallelogram, and that the height of the triangle is also the same as the height of the parallelogram (Figure 8). Students can thus see that the area of the parallelogram is two times the area of the triangle with the same base and the same height. The teacher can also ask students to express the relation between the area of the triangle and the

parallelogram using algebraic notation, $A = 2 \times \dfrac{1}{2} bh = bh$.

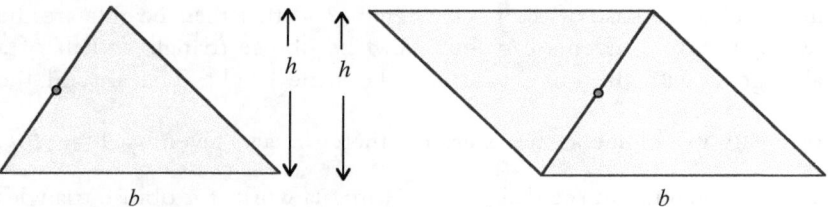

Figure 8. A parallelogram with the same base and same height.

Interactive figure 6 shows another connection between the area of a triangle and a parallelogram. The hinge is at the midpoint of a side of the triangle, and a small triangle is formed by a segment through the midpoint parallel to the base. Students can be guided to see and justify why the rotated small triangle together with the part of the triangle that did not move form a parallelogram, the area of the triangle is the same as the area of the parallelogram, the base of the triangle is the same as the base of the parallelogram, and the height of the parallelogram is half the height of the original triangle. If we let students express the height of the parallelogram as h, they will express the height of the triangle as $2h$ (Figure 9). They can then express the area of the triangle algebraically, and simplify the expression to

obtain a formula for the area of a parallelogram, $A = \dfrac{1}{2}b \times 2h = bh$.

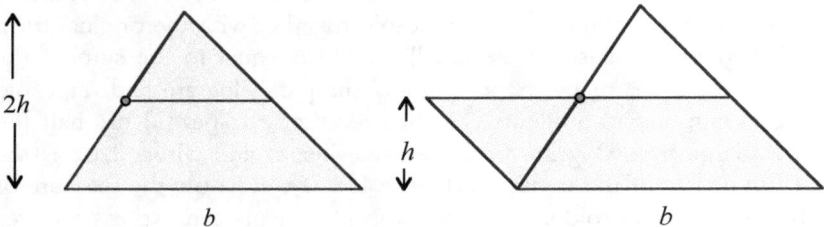

Figure 9. A parallelogram with half the height.

The approach for the area formula for triangles developed in interactive figures 1 through 6 works for right triangles and triangles with the altitude inside the triangle. Once we know the formula for the area of a parallelogram, we can find that the formula for the area of a triangle works for all kinds of triangles, even when the obtuse angle is at the base, that is, when the altitude lies in the triangle's exterior. In interactive figure 7 the hinge is at the midpoint of the longest side of the triangle (the side opposite the obtuse angle). Students need to make explicit that the rotated triangle together with the original form a parallelogram, that the base of the triangle is the same as the base of the parallelogram, and that their heights are the same (Figure 10). Students can thus view the obtuse triangle as half of a parallelogram with the same base and the same height. The area of the triangle with its altitude in the exterior is therefore also given by $A = \dfrac{1}{2}bh$.

Of course students can see that the area formula works for obtuse triangles using other methods, for example, expressing the area of the obtuse triangle as the difference of the areas of two right triangles.

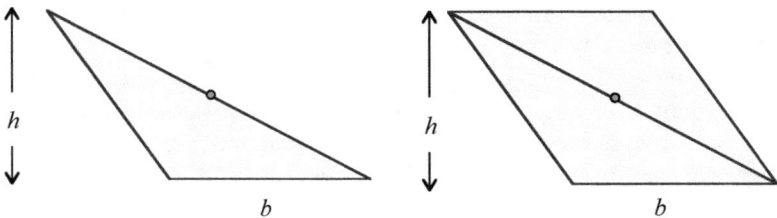

Figure 10. Triangle with altitude outside.

Trapezoid and parallelogram

The formula for the area of a trapezoid can be related to the formula for the area of a parallelogram. In interactive figure 8 the hinge is at the midpoint of one of the legs of the trapezoid. The teacher can guide students to see and justify why the rotated trapezoid together with the original form a parallelogram, the base of the parallelogram is equal to the sum of the lengths of the bases of the trapezoid, and the parallelogram and trapezoid have the same height. Students can thus see that a trapezoid has half the area of a large parallelogram with the same height and whose base is the sum of the bases of the trapezoid (Figure 11). If we denote the measure of one base of the trapezoid as a, the other as b, students can express the area of the large parallelogram as $(a+b)h$. Students can therefore write the area of the trapezoid as $A = \dfrac{1}{2}(a+b)h$.

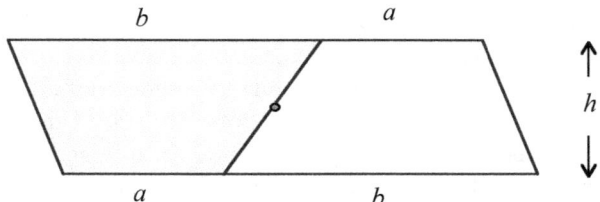

Figure 11. Trapezoid and parallelogram same height.

Students can also see that the segment joining the midpoints of the legs of a trapezoid is parallel to the bases and has a length equal to the arithmetic average of the two bases, $\dfrac{a+b}{2}$ (interactive figure 9), because two times this segment is equal to the sum of the bases. By writing the formula as $A = \dfrac{a+b}{2}h$, students can see another interpretation for the area of the trapezoid as the product of the average of the bases times the height (Figure 12).

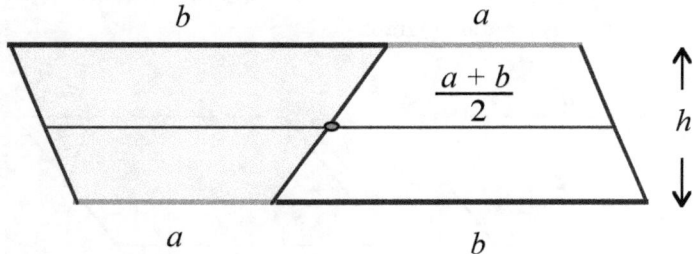

Figure 12. The average of the bases.

In interactive figure 10 teachers can direct the attention of the student to the distance from the segment through the midpoints of the legs to each of the bases. Students can realize and justify why the line is equidistant from both bases and why the distance is half the height of the original trapezoid. Thus, when the upper part of the trapezoid is rotated, a parallelogram is formed (Figure 13). The teacher can ask questions so that students make explicit that indeed the new shape is a parallelogram, that its height is half the height of the original trapezoid, and that the base of the long parallelogram is the sum of the bases of the trapezoid. Students can state that a trapezoid has the same area as a parallelogram with half the height and whose base is equal to the sum of the bases of the trapezoid,

$$A = \frac{h}{2}(a+b).$$

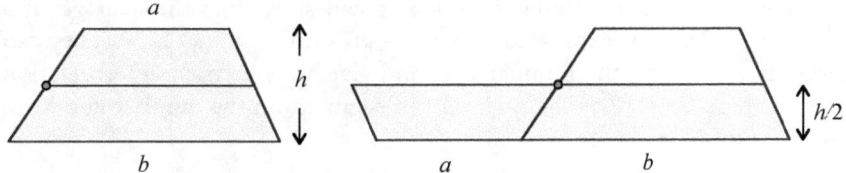

Figure 13. Parallelogram with half the height.

We have thus three equivalent formulas for the area of the trapezoid,

$\frac{1}{2}(a+b)h = \frac{a+b}{2}h = (a+b)\frac{h}{2};$ each one has a different geometrical interpretation. Students should be able to explain when we use associative and commutative properties to transform one formula algebraically into another.

We can also transform a trapezoid into a parallelogram by rotating one small triangular part. In interactive figure 11 the hinge is at the midpoint of one of the legs of the trapezoid. The small triangle has one side parallel to the other leg of the trapezoid. The teacher can ask students to justify that indeed the new shape is a parallelogram.

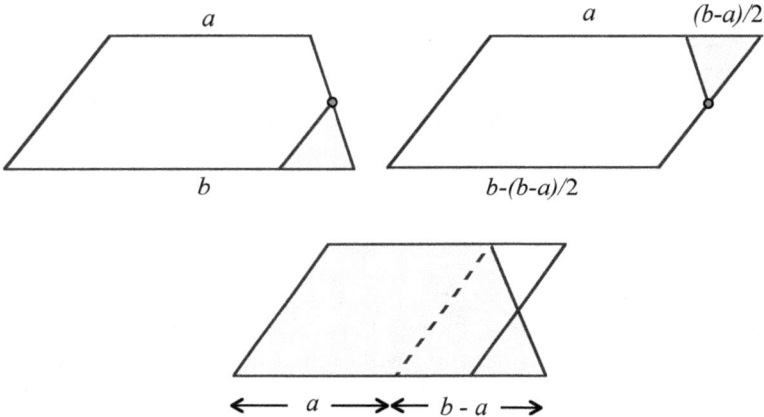

Figure 14. Base of parallelogram is the average of the bases of trapezoid

Finding the length of the base of this parallelogram may not be immediately obvious, however. Students might observe that the base is equal to the segment through the hinge and parallel to the bases of the trapezoid. Alternatively, if the length of the larger base is b and the shorter base is a, students can see that the length of the base of the parallelogram in Figure 14 is $(a+b)/2$ in the following way. Let a and b be two numbers with $a \le b$, and let $b - a$ be their difference. Students can show that the average of a and b, $(a+b)/2$ can be expressed as $a + (b - a)/2$, or as $b - (b - a)/2$. They can represent a and b on the number line, and give a geometrical interpretation of the average $(a+b)/2$ as the midpoint of segment on the number line from a to b (Figure 15).

In either case, as they saw before in interactive figure 12, the length of this segment is the average of the lengths of the bases. Thus students can see in another way that a trapezoid has the same area as a parallelogram with the same height and whose base is the average of the bases of the trapezoid, $A = h\dfrac{a+b}{2}$.

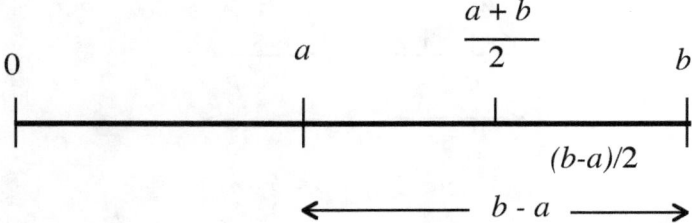

Figure 15. The average as the midpoint of a and b.

Trapezoid and rectangle

A trapezoid can also be transformed into a rectangle by rotating two small triangles. In interactive figure 12 the hinges are at the midpoints of the legs of the trapezoid. Students can remember that the line segment connecting these midpoints will be parallel to the bases of the trapezoid, and its length will be the average of the bases. Figure 16 shows a trapezoid that has the same area as a rectangle with the same height, and whose base is the average of the bases of the trapezoid. Students can also verify algebraically that indeed the base of the rectangle is the average of the bases of the trapezoid in the following way. Let q be the length of the projection of one leg onto the longer base, and p be the length of the projection of the other leg (Figure 17). Then $q + p = b - a$. The bases of the small triangles are $q/2$ and $p/2$. When the triangles are rotated up (Figure 18), students can se that the base of the rectangle is $a + q/2 + p/2 = a + (q + p)/2 = a + (b - a)/2 = (a + b)/2$. Here again the teacher may want to help students recognize the use of the distributive property in the first equation, and the associative and commutative properties in the second.

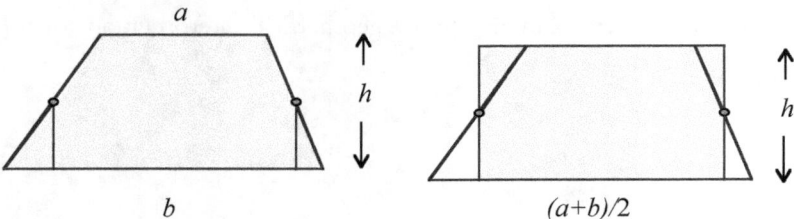

Figure 16. The base of the rectangle is the average of the bases.

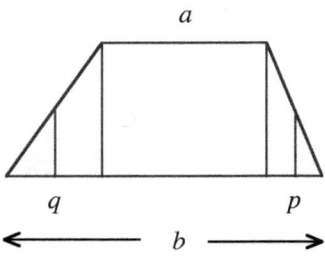

Figure 17. $q + p = b - a$

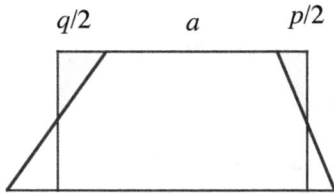

Figure 18. Base of the rectangle.

Kite

In a kite, the two diagonals are perpendicular. In interactive figure 13 the hinges are at the midpoints of the sides of the kite. Four right triangles form the kite. The four rotated triangles together with the original kite form a rectangle (why?). The teacher can ask questions to focus the attention of the students on the relation between the lengths of the diagonals of the kite and the base and height of the rectangle. Students can see and justify why the area of a kite is half the area of a rectangle with height equal to one diagonal d of the kite and base equal to the other diagonal c of the kite (Figure 19). The area of the kite is thus $A = \frac{1}{2}dc$. By focusing on one half of the rectangle, students can also find a geometrical interpretation for the expression $A = d\frac{c}{2}$.

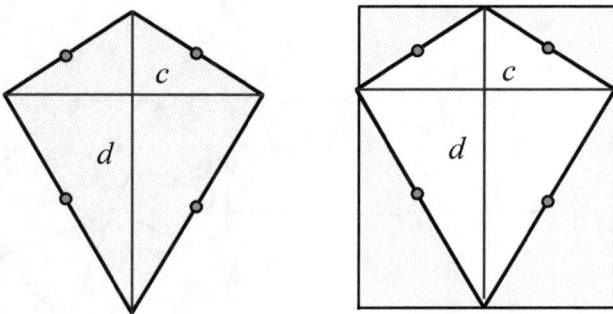

Figure 19. Kite and rectangle.

Regular polygons

Interactive figure 14 shows a connection between the area of a parallelogram and the area of a regular polygon with an even number of sides. Triangles of two different colors each form half the regular polygon. (Figure 20). Therefore the length of the parallelogram's base is the sum of the lengths of half the number of sides of the polygon and the base of the parallelogram is equal to half the perimeter of the regular polygon, that is, $b = \frac{1}{2}p$. The apothem a of a regular polygon is the distance from its center to the midpoint of one of its sides. We can see that the height of the parallelogram is equal to the apothem a of the regular polygon. Thus, the area of the parallelogram is given by $A = \frac{1}{2}pa$, or $A = \frac{pa}{2}$. The area of the regular polygon will thus be $A = \frac{pa}{2}$.

In the case of a regular polygon with an odd number of sides, a rectangle is formed (Figure 21), and the reasoning is the same to show that the area of a regular polygon with an odd number of sides is also $A = \frac{pa}{2}$.

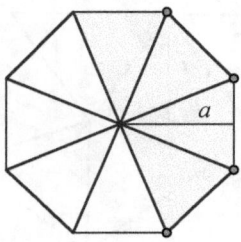

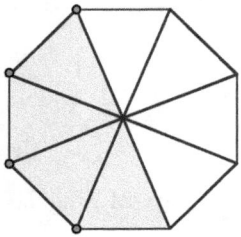

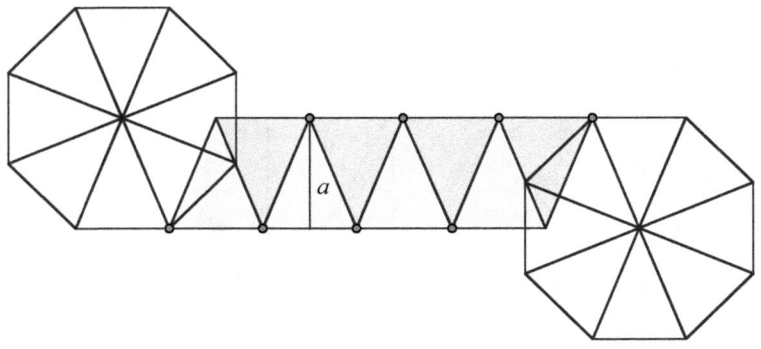

Figure 20. Regular polygon (even number of sides) and parallelogram.

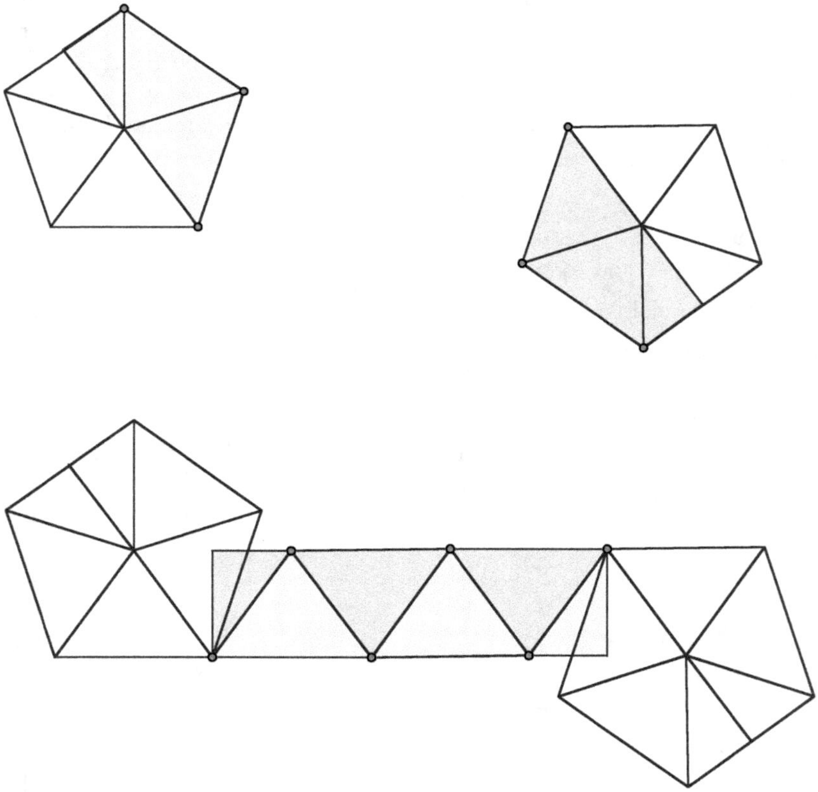

Figure 21. Regular polygon (odd number of sides) and rectangle.

Concluding remarks

These interactive figures, with appropriate guidance from the teacher, give students the opportunity to connect the formulas for the areas of the rectangle, right triangle, arbitrary triangle, parallelogram, trapezoid, kite, and regular polygon. Rather than having a collection of isolated formulas, students can see how the formulas form a network of relations and how some of them may be derived from other formulas or from special cases. The interactive figures can also serve as an experimental laboratory to facilitate the discovery of properties of shapes and transformations to be proved later, and to provide opportunities for conjecturing and practicing deductive thinking as students justify their observations. The activities also can be used to develop the algebraic thinking of students. It is important for students to give meaning to algebraic expressions, and to see why certain operations on algebraic expressions will indeed give equivalent expressions. When students are still developing their understanding of algebraic operations, contexts arising outside algebra can help them develop meanings. They may need to form specific images for the algebraic expressions, and be able to explain them verbally or by using pictorial representations. Formulas for areas can thus serve also as a vehicle to give meaning to and practice for algebraic operations.

19 FORMING A CADRE OF MATHEMATICS MENTORS[19]

We describe a project to form a cadre of mentors for the 5th and 6th grades in Glendale Elementary School District in Arizona. The project to prepare mentors was funded through a couple of grants and was part of a partnership formed by district teachers and curriculum facilitators, and university faculty that spanned several years.. This partnership started with the commitment of a few individuals several years before. As a classroom teacher, the second author took several graduate mathematics methods courses, and then a change in position in her district enabled her to see the need for district wide change and improvement in the instruction of mathematics. The first author had offered graduate courses at schools in the district and several professional development sessions on-site at a neighboring school.

Glendale Elementary District faced many challenges in the teaching of mathematics. About 32% of the teachers had three years or less of experience. A large part of the mathematics teachers in grades 5 - 8 were certified as elementary teachers, and needed more mathematical and pedagogical content knowledge. There was a great pressure for improving the performance of the district's schools. Glendale schools' performance in norm reference test in mathematics was consistently below that of the state as a whole. A large percentage of students did not meet the state standards

---

[19] Flores, A. and Thomas, C. A. (2009). Forming a cadre of mathematics mentors. In G. Zimmermann (Ed.), *Empowering the mentor of the experienced mathematics teacher*, pp. 49-50. Reston, VA: National Council of Teachers of Mathematics. Copyright National Council of Teachers of Mathematics. Used by permission.

in mathematics. Seven schools were labeled as underperforming. There was a wide variety of methods, approaches, and textbooks used in the different schools. One of the goals of the district was to give teachers the necessary support in professional development to implement methods to teach mathematics for understanding. The strategy to make these profound changes district wide was to form a cadre of mentor teachers.

Teachers participated in two different phases of the project, Spring course and Summer session. In the Spring, 15 teachers took a graduate course in mathematics methods for credit, focusing on teaching middle grades. The course was offered at one of the schools of the district, met for 3 hours each week during 14 weeks. Special attention was given to topics that are crucial in the middle grades, such as number sense and place value; meanings of operations and operations with integers; fractions, rational numbers, and decimals; transition to algebra; ratio, proportion, and similarity; functions and graphs.

The majority of the these teachers, together with an additional number of teachers participated in the summer session, for a total of 26 teachers. The Summer component consisted of 12 full day workshops. During the morning, teachers observed lessons with students. The district expert modeled some of the activities with students. These lessons exemplified best practice. For example, one of the sessions included active participation of the students in using pattern blocks to create designs and discuss the different kinds of symmetries (mirror, rotational) present in them. Students identified lines of symmetry and angles of rotation that would leave the design unchanged. Other activities for students included keeping math journals, and mathematical games.

Teacher participants did not have a passive role observing. They interacted with the children, made their own word problems and provided learning opportunities for the children. Then, all teachers together with the curriculum facilitator debriefed the sessions with students in the morning. Teachers analyzed what was effective and why; what could be improved and how; and also how to make changes that would better fit their own styles and the needs of their own students. In the afternoon, teachers formed small groups to discuss the focal question of the day and other issues. In the second part of the afternoon teachers also engaged in activities using the same hands-on materials and topics as the children but that demanded a deeper level of understanding. For example, teachers used the same pattern blocks that had been used by the children earlier, together with hinged mirrors used as simple kaleidoscopes. Teachers identified angles that would give back each of the selected pattern blocks when the vertex of the hinged mirror was placed at the center. Ideas such as factors and multiples and compositions of reflections were discussed.

There was continued support for mentor teachers during the school year. Two four-day sessions were conducted in the next Fall and Spring following a similar format as in the Summer. In addition, mentor teachers met monthly. The mentors improved their own practice and provided leadership and support to other teachers in the district. Teachers implemented standards-based curricula in the middle grades. Participants learned teaching methods that address national and state standards, and that are based on research and best practice. These changes have been integrated into daily practice and have been sustained over time. Teachers have strengthened their understanding of content and of how students learn mathematics best. They are also more reflective about their own teaching. Participants were enabled to work with other teachers to disseminate the changes in their own schools. They have contributed to the district's overall efforts as part of the leadership teams. By building participants profound understanding of the mathematics they teach, as well as expanding their pedagogical content knowledge they have contributed to sustain change in the district. Participants worked with other teachers to disseminate the changes in their own schools. They have contributed to the district's overall efforts as part of the leadership teams.

20 UNDERSTANDING RIGID GEOMETRIC TRANSFORMATIONS: JEFF'S LEARNING PATH FOR TRANSLATION[20]

Abstract

This article describes the development of knowledge and understanding of translations of Jeff, a prospective elementary teacher, during a teaching experiment that also included other rigid transformations. His initial conceptions of translations and other rigid transformations were characterized as undefined motion of a single object. He conceived transformations as movement and showed no indication about what defines a transformation. The results of the study indicate that the development of his thinking about translations and other rigid transformations followed an order of 1) Transformations as undefined motions of a single object, 2) Transformations as defined motions of a single object, and 3) Transformations as defined motions of all points on the plane. The case of Jeff is part of a bigger study that included four prospective teachers and analyzed their development in understanding of rigid transformations. The other participants also showed a similar evolution.

1.   Introduction

There has been a growing emphasis on teaching geometry in the last few decades both at the elementary and the secondary level. The National

---

[20] Yanik, H. B. and Flores, A. (2009). Understanding rigid geometric transformations: Jeff's learning path for translation. *Journal of Mathematical Behavior*, *28*(1), 41-57. Reprinted by permission.

Council of Teachers of Mathematics (NCTM) (1989, 2000) has recommended that geometry be an integral part of the school mathematics curriculum at all levels. Transformational geometry is one of the topics that has been advocated as an important part of the K-12 geometry curriculum. According to NCTM's *Principles and Standards for School Mathematics* (2000), "Instructional programs from pre-kindergarten through grade 12 should enable all students to apply transformations and use symmetry to analyze mathematical situations" (p. 41). To accomplish this goal, teachers should have strong content knowledge background to create classroom environments where students develop reasoning and justification skills (Parsons, 1993; Leinhardt & Smith, 1985).

However, several research studies (e.g. Mayberry, 1983; Parsons, 1993; Ma, 1999) have documented that both prospective and in-service teachers do not always have a profound understanding of the mathematics they are supposed to teach. Particularly, studies on geometric transformations have shown that pre-service elementary school teachers have a lack of understanding of geometry, in particular, comprehending concepts of transformations including translations, reflections, rotations and compositions of transformations of those types (Law, 1991; Edwards & Zaskis, 1993; Desmond, 1997; Harper, 2002).

Although past research focused on describing prospective teachers' difficulties in understanding of geometric transformations, few studies have been conducted to research the growth of prospective teachers' knowledge and understanding of rigid transformations over time. More research needs to be conducted, not only to identify what possible understandings pre-service teachers hold, but also to examine what critical factors and relationships are necessary to establish a progression into more sophisticated and advanced ways to think about transformations.

2. Framework for the study

First, we list the mathematical definitions of terms used in this article. Next, we examine the research on understanding geometric transformations. Lastly, we describe and analyze one of the participants' (Jeff) growth in knowledge and the understanding of rigid transformations and how his thinking evolved towards seeing transformations as mappings of the plane onto itself.

This case study of Jeff is a part of a larger study which included four prospective elementary teachers which explored the following questions:

- What subject-matter content knowledge do pre-service elementary school teachers have about geometric transformations (i.e., translations, reflections, and rotations)?
- How do pre-service teachers' understanding of transformations as motions evolve to seeing transformations as mappings of the

plane?
- What works in developing pre-service teachers' understanding of rigid transformations? How? When? and, Why does it work?
- What kinds of changes occur in prospective elementary teachers' knowledge of geometric transformations?

2.1. Definition of terms

This section provides the definitions of terms that refer to the specific meanings used in this study. The term *transformation* refers to the rigid transformations unless it is explained differently.

A rigid transformation. A transformation that preserves relative distances and angles of all points in the plane.

Domain. The term *domain* refers to a specific meaning in geometric transformations, considered as functions of the plane. In geometric transformations, the domain is considered as all the points in the plane (Hollebrands, 2003). That means it is not limited to a figure or a point. When translating a figure in the plane, all points in the plane need to be considered.

Input and output. In functions, an *input* can be any value based on the domain defined for the function (e.g., the domain of function f). For transformations, since the domain is defined as all points in the plane, all points in the plane must be considered as input rather than a point, or a line, or a geometric figure alone. *Output* is the result of applying the transformation.

Parameters. Parameters for transformations include *translation vectors, reflection lines, points for centers of rotations*, and *measures of angles of rotation* (Hollebrands, 2003).

Plane. The plane consists of an infinite number of distinct points and geometric figures that are subsets of points of the plane rather than separate entities.

Motion and mapping understanding of transformations. A person who has a motion understanding of transformations may conceive the plane as a background and geometrical objects could be manipulated *on* it. According to Edwards (2003), in this understanding one may consider transformations as "physical motions of geometric figures on top of the plane" (p. 8). For a mapping understanding, on the other hand, one might think that the plane consists of an infinite number of distinct points and geometric figures are subsets of points of the plane rather than separate entities (Edwards, 2003). In other words, in the mapping understanding one maps all points in the plane to other points in the plane rather than removing images/points from their original locations to locate them to different locations. In this understanding, while motion may be still a useful way to look at transformations in some cases, it is not an essential aspect of the mapping. Thinking of transformations as mappings allows us to think about transformations beyond motions.

2.2. Research on prospective elementary teachers' understanding of geometric transformations

The purpose of this section is to review the literature to clarify concepts and summarize relevant research to understand prospective elementary teachers' growth in knowledge and understanding of rigid geometrical transformations.

Studies have documented prospective elementary teachers' difficulties in understanding various concepts related to transformations (Desmond, 1997; Edwards & Zaskis, 1993; Harper, 2002; Law, 1991).

Law (1991) investigated how pre-service elementary school teachers developed and learned the concept of translation, reflection and rotation. Law interviewed eighteen pre-service elementary teachers who took a geometry course in which geometric transformations were introduced in classroom lectures. During the interviews prospective teachers were asked to define each type of transformation and provide examples to show their thinking. Law pointed out that pre-service elementary school teachers had difficulties in giving the definitions of transformations whether in a concrete way or using abstract terms. He stated that "They either gave a wrong description of those transformations, or could not manipulate points in order to show what the action of flip, turn, or slide is" (p. 48). He added "We need to ask some more questions in order to investigate what the definitions are in their minds" (p. 72).

In their study, Edwards and Zaskis (1993) found that prospective elementary teachers "did not initially think of the transformations as mathematically-general operations which required specification of inputs but instead as particular actions, each with given 'default' or prototypic parameters" (p. 130). The researchers asked fourteen pre-service elementary teachers to rotate and flip various objects (e.g. a book, a pen, a flag and so on) and the students carried out the requests without asking for explicit instructions except in two cases. In those examples, both involving rotation, one student wanted to know how many times to turn the object, and the other student wanted to know in what direction to rotate it. The results indicated that the students' primary images of "rotations and reflections feature center points generally located at the visual center of the figures or in a few cases, at special 'vertex.' Mirror lines either cut the objects in half or a salient 'long edge' of the shape" (p. 133).

Pre-service elementary teachers also have difficulties in determining: (1) the correct transformation and motion attributes to move an object from one point to another; and, (2) the results of transformations involving multiple combinations of figures (Desmond, 1997). Desmond found that only 17% of the 83 prospective elementary teachers identified a translation with the correct distance and direction; 24% of them recognized a reflection with the correct line of reflection; and 18% of the participants identified a rotation with the correct point, direction and degrees of rotation on the item as shown in Figure 1.

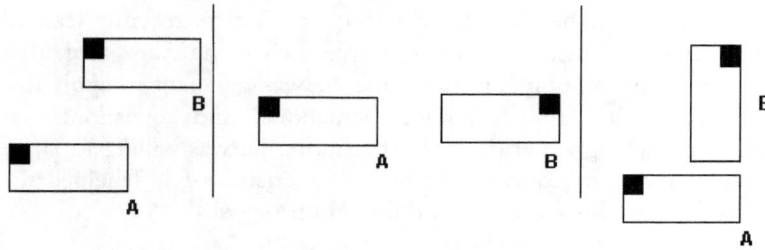

Fig 1. Transformations-Translation, Reflection, and Rotation (Desmond, 1997, p. 71).

Harper (2002) found that prospective elementary teachers' initial vocabulary for using transformations was very primitive. They also had problems in "constructing translations vectors to map a figure to its image, as well as using translation vectors to construct the image of a figure" (p. 104). In addition, students were weak in understanding rotations and identifying the center and angle of rotation of a figure and its image. Some of the students were weak in determining a single rotation that would match a figure with its image.

3. Methodology

3.1. Setting and Participants

This article is based on the first author's doctoral dissertation (Yanik, 2006) done under the supervision of the second author. Whenever we use the term "the researcher" in this article it refers to the first author. The study was conducted at a large metropolitan public research university in the southwest of the United States. A total of four prospective elementary teachers from the Curriculum and Instruction Department voluntarily participated in this study. Two of the participants were majoring in elementary education and in their last year of an undergraduate program. The other two participants were attending a master's program in mathematics education.

Before the study began, three of the four participants took a one-hour tutoring session on the Geometer's Sketchpad to learn the basics of this program. The fourth participant, Jeff, was enrolled in a technology class which included learning the Geometer's Sketchpad at the time this teaching experiment was conducted.

Jeff was chosen as a case for this study because he moved further ahead in the development of his thinking about transformations and because he was a very reflective person, who was ready to accept challenges and was able to share his thoughts explicitly. One striking character of Jeff was that he always questioned what he thought was true or what he recently learned during the teaching episodes. For example, when he was told that transformations apply all points in the plane, he was reluctant to apply this "new knowledge" to transformation situations and questioned it throughout the teaching experiment. Furthermore, he was willing to share his thoughts without any concern. These characteristics of Jeff helped the researcher understand his reasoning and thought processes.

3.2. Teaching Experiment
Individual teaching experiment methods (Steffe and Thompson, 2000) were utilized in this study to investigate the development of pre-service elementary teachers' subject-matter knowledge and understanding of geometric transformations. The teaching experiment included three phases: task-based clinical interviews (pre and post), teaching episodes, and data analysis.

3.2.1. Task-based Clinical Interviews
At the beginning of the study the researcher conducted a semi-structured task-based clinical interview with each participant to assess four pre-service elementary teachers' pre-existing knowledge of rigid geometric transformations. Interview questions required prospective teachers to give descriptions, provide examples and non-examples of transformations, and perform and identify transformations. During these interviews, the researcher had no intention of advancing the participants' actual knowledge. The final clinical interviews aimed at identifying the current conceptual constructions of rigid geometric transformations of four future teachers. The tasks that were used for these interviews were designed to allow students to reflect on what they had experienced in the teaching episodes, and to help the researcher identify participants' knowledge of geometric transformations.

3.2.2. Teaching Episodes
The purpose of the teaching episodes was to create situations that would encourage participants to modify their thinking and so that the researcher

would be able to gather in depth understanding of prospective teachers' changing conceptions of geometric transformations.

For eight weeks, the researcher met with each participant one at a time once or twice a week for two hours for individual (one-on-one) instructional sessions in a math lab. The participants spent at least three sessions working on each transformation. Based on their level of understanding, the researcher added further teaching episodes to get a complete picture of their conceptions of transformations. All participants were given similar tasks, but probing questions were varied in complexity, depending on each participant's level of understanding determined by the first clinical interviews. The tasks were not too simple or difficult but a little bit above the participant's level. In this way, the participant would need to push his/her limits to learn new concepts.

During the teaching episode, the first author of this article served as the instructor, and continually interacted with the participants and analyzed their explanations. After each teaching episode the researcher coded and analyzed the video records of subjects interactions with the given tasks. He looked for patterns and described the subjects' development of understanding of transformations. These descriptions were the base for developing the next instructional session and eventually the hypothetical learning trajectory of each participant (see data analysis section below).

For the teaching episodes several tasks related to transformations (i.e., translations, reflections, and rotations) were designed to reveal the participants' understanding of geometric transformations. Table 1 lists both the types of tasks used throughout the study and the timeline for Jeff for translations.

Table 1
Task types and Jeff's study schedule for translations

Task Types	The task is about	Time Line
Type 1	describing and performing a translation	Week 1
Type 2	understanding the parameter and its relation with pre-image and image points	Week 1, 2, &3
Type 3	identifying the parameter	Week 3, 4
Type 4	understanding fixed points.	Week 3

Type 5	understanding the domain of translation	Week 3 & 4
Type 6	performing compositions of translations	Week 4
Type 7	describing single and multiple translations using a function notation.	Week 4

The first type of task was about describing and performing a transformation. The reason behind this task was to see whether the participant would describe and successfully execute a transformation or not. For example, if the participant could not perform a translation, the researcher aimed to understand what would be the possible reasons and missing concepts.

The second type of task was about understanding the parameters and their relations with pre-image and image points. Since the parameters define transformations, it is crucial to understand the relationships between the parameter and the related transformation. For example, the length and the direction of a vector determine where the points in the plane are going to be mapped. The line of reflection defines a reflection, and the center of rotation and angle of rotation are the parameters that define rotations.

The third type of task was about identifying a parameter. For example, for two given figures (pre-image and image), participants were asked to find out the parameter that would determine the given transformation. This task was helpful to understand whether the participants really understood the concept of parameter.

The fourth type of task was about understanding fixed points. One of the important concepts that participants needed to know was that points could be mapped onto themselves as a result of a transformation. Therefore, participants were asked whether it could be possible to map a pre-image point onto itself under a transformation.

The fifth type of task that was used in this study was about the domain of transformations. Understanding the domain as all points in the plane was another important concept for participants to understand rigid transformations. For example, one needs to apply a transformation to all points in the plane rather than a single geometrical figure. Therefore, participants were provided multiple geometrical figures and were asked whether they would consider all points in the plane or just one of the figures to be transformed.

The participants were asked to perform compositions of transformations. This question was informative in terms of understanding whether participants would see these transformations as two separate processes or whether they could be represented by a single transformation.

The participants were also asked to describe single and multiple transformations using a function notation. Representing a transformation with a function notation requires more abstract thinking and is crucial for understanding transformations as one-to-one mapping of the points of the plane (Flanagan, 2001). We should note here that during the teaching experiment a sample function notation for a single translation was given and discussed with the participants before they were asked to represent each translation with the notation.

Regarding the teaching aspect of teaching episodes, the researcher tried to create a learning cycle for the participants which included three phases, "an exploration phase, concept introduction phase, and concept application phase" (Engelhardt, Corpuz, Ozimek, and Rebello, 2003, p. 2). In the first stage, the researcher provided each participant a learning space in which participants explored the tasks shown in Table 1 using the dynamic geometry software, the Geometer's Sketchpad. For example, for type-one questions each participant was initially provided a semi-structured task (e.g., Can you translate the figure given on the computer screen using the Geometer's Sketchpad?). The researcher asked the participant to predict the result of the translation before he/she actually executed the translation. Then the participant demonstrated the result of translation by performing it using the Geometer's Sketchpad and was asked to explain what he/she realized or expected to happen in the demonstration. Each participant was allowed to repeat the activity if it was necessary. In the following phase (the concept introduction phase) "an explanation of the observations that were performed in the exploratory phase is given a name and further refined" (Engelhardt et al, 2003, p.2). For example, for type-two task, the researcher hoped that the participants would understand the role of parameter (i.e., the vector) and its relation with pre-image and image points. In order to help the participants grasp the vector concept, the researcher asked them to reflect on the effect of changing the magnitude, direction and the location of the vector on the result of the translation. Furthermore, probing questions directed participants' attention on the *effect* of translation that is what has been changed and what has not been changed as a result of altering the features of the vector. In the concept application phase, the participants "apply the concept they explored and later named to new situations" (Engelhardt et al., 2003, p. 2). For instance, for type-five questions the participants were introduced to the domain as all points in the plane and tried to apply this knowledge to determine the result of translation for given multiple figures.

3.2.3. Data analysis

The data collected through clinical interviews and teaching episodes were analyzed using ongoing and retrospective analyses. Ongoing analysis took place during and between teaching episodes and a retrospective analysis focused on the cumulative episodes.

During the ongoing analysis phase, the researcher tried to understand the participant's way of thinking. During each teaching episode, the researcher asked several probing questions to make the participant's understanding explicit. After each teaching episode, the researcher coded and analyzed the video records of students' interactions to the given tasks. The main purpose was to find patterns and create descriptions of the students' mathematical knowledge development over time. These descriptions helped the researcher portray the students' current understanding of knowledge of transformations and would be the base for both developing the next instructional session and the hypothetical learning trajectory (Simon, 1995) of participants. In the case of emerging different trajectories, the researcher noted each student's similarities and dissimilarities and developed tasks for the individual to help him/her progress in learning. During ongoing analyses the researcher tested his initial hypotheses and generated new conjectures to be tested in the following teaching episodes.

Retrospective analyses involved careful analysis of the videotapes after completion of each teaching episode and the teaching experiment. This provided the researcher an opportunity to activate the records of past experiences with the subjects. In this way, the researcher was able to make a historical analysis of the participants' mathematics both retrospectively and prospectively. The main purpose of this analysis was to gather in-depth information about development in the students' knowledge of rigid transformations. During this exploration, the researcher searched for answers to questions, such as "What works in developing pre-service teachers' understanding of transformations as motions to see transformations as mappings of the plane?" "How, when, and why does it work?" In researching these questions, the researcher developed hypotheses and looked for the data to support or challenge his assumptions. The coding process helped the researcher organize his data, and the theoretical model (see Figure 2) of Jeff's growing understanding of translation became apparent eventually.

4. Analysis

4.1. A case study of Jeff

Jeff was a 25 year-old student in the first year of a master's program in mathematics education. He graduated with a bachelor degree in elementary education from a southwestern university just before he started the program at another southwestern university. His only direct teaching experience was student teaching in a first grade classroom, as well as observing numerous classes.

According to Jeff, his background experience with mathematics is based on his love for numbers since an early age and a personal search for knowledge. He took four math courses in college and was interested in all of them. His highest math course was calculus with matrices. "I love mathematics. Ever since I was young, my father would write out lists of problems for me to compute and I always enjoyed them," he remembered. The last formal instruction Jeff received in geometric transformations before these teaching experiments was during high school. He remembers problem solving and estimation of geometric transformations from high school. He said, "My experiences were focused on the basic transformations, slides, rotations, and dilations. Just completing problems, and a lot of estimation." He had been using the Geometer's Sketchpad for a few weeks in a *Teaching Math with Technology* class.

4.2. Jeff's learning path for translation

This section presents Jeff's learning path for translation. Based on the clinical interviews and the teaching episodes conducted throughout the study, a theoretical model was generated to describe a hypothetical development of Jeff's understanding of rigid transformations. The model describes Jeff's developing ideas of translations as he moved from an understanding of transformations as undefined motion of single objects toward an understanding of transformations as mapping of the plane onto itself (see Fig.2). Although Jeff did not quite reach the mapping understanding of transformations, the results revealed both a deficiency in initial understanding and opportunities to enrich those understandings. The following sections describe and analyze the development of Jeff's understanding of translations.

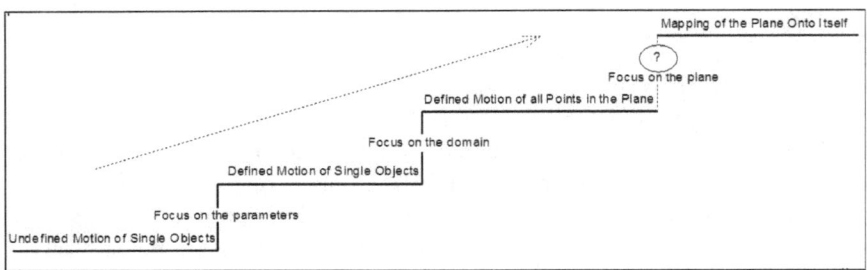

Fig 2. Jeff's hypothetical learning trajectory.

4.2.1. *Translations as undefined motions of a single object*

This section presents an analysis of the first set of clinical interviews dedicated to understanding Jeff's initial understandings of rigid translations. The first clinical interviews revealed that Jeff had a motion conception of translations. Jeff was thinking that translation was a movement rather than mapping of the plane onto itself. He considered the domain as a single object and translations applied to only one object rather than all points in the plane.

Jeff also had limited understanding of the parameter, the vector. He did not know how to use the vector and whether it was necessary for a translation or not. Jeff could not explain what defines a translation. It seemed that translation was an undefined motion of an object on the plane, which could be any direction or distance. Detailed examples appear below.

During the initial clinical interview, Jeff interpreted a translation as a motion of a single object without changing its shape, size, or orientation. Jeff's explanation ("the word *trans* makes me think that it [a translation] was movement") and examples supported this conception. The following excerpts illustrate how he was thinking about translations during the interview.

R: Could you please give me an example of a translation?

J: For my definition, translation is when an object is just moved and nothing is changed about the object. It is the exact same in a size and angle [orientation]. If I have this object here, I would translate it just go right over there [he took a hexagon from the materials and moved it on the paper]. It is just moved (see Fig. 2).

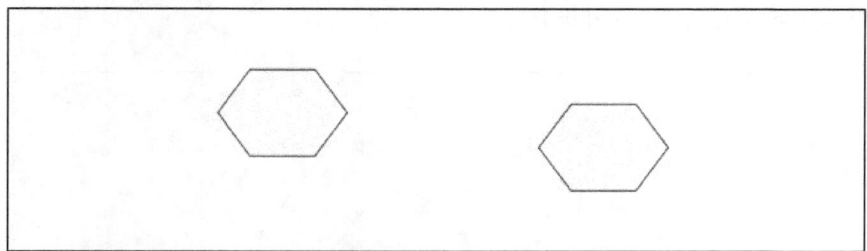

Fig. 2. Jeff's example of translation.

R: How about a non-example of a translation?

J: Non-example would be... Maybe you move and then you turn it something [he took a hexagon and slid it and turned it] on the paper and so that the coordinating points are no longer the same. Like this is A here and A' would be there. Instead the non-example A' would be over here. That would be more likely rotation (see Fig. 3).

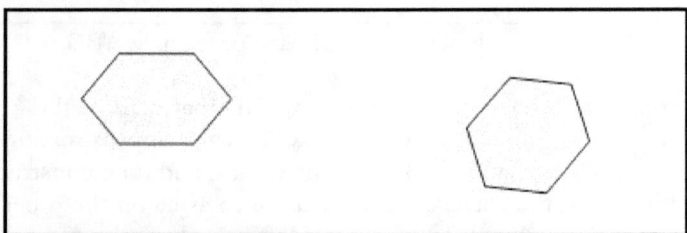

Fig. 3. Jeff's non-example of a translation.

R: Would you define a translation?

J: I would define translation as moving of an object from one point to another without any changes in the object and all the points can be related back to the first object. They are all on the same position.

Although Jeff was reasoning from the motion perspective of translations, he also showed some understanding of rigid transformations. He knew that the attributes of shapes (e.g., shape, size, angle measures, and orientation) would be preserved under a translation.

4.2.1.1. Jeff's initial conceptions about the parameter
The first clinical interviews also revealed that Jeff had an incomplete understanding of vectors. In a translation task (see Fig. 4), Jeff explained that the image of the triangle was going to cross "the line" [vector $\vec{V}$]. He

drew the image as shown in Fig. 5.

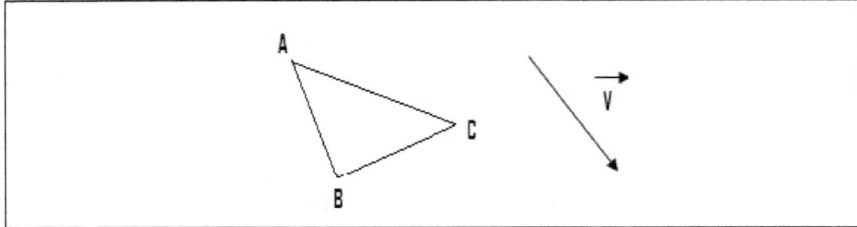

Fig. 4. First clinical interview task.

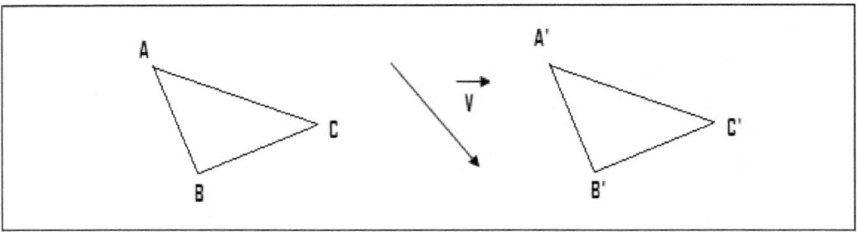

Fig. 5. Jeff's drawing of the image of triangle ABC.

Jeff was considering the vector as if it was a line that marked the middle of the translation. He initially tried to make a visual approximation of the distance between the triangle ABC and the vector, and then constructed the translated image of the triangle a similar distance away on the other side of the vector. The following dialogue reveals Jeff's thinking about the vector.

R: Could you please tell me how you determined where to place the triangle ABC?

J: I determined where to place the triangle relative to the line [vector $\vec{V}$] that was given. I mean I chose to put it [A'] right here based on the distance from this point [A] to this line [vector $\vec{V}$]. I chose to put it [A'] a similar distance on the other side of the line. I did not put it over here on the edge of the paper. I put it relatively close to the line that was being translated.

R: Ok. Can you show me the distance again?

J: I was thinking that if it is right here [triangle ABC] and this is the line [vector $\vec{V}$] right here. Then I translate it over that line. Then I don't think it would be good if I put C' here [edge of the paper] even though it would still be translation. I just think that this would be the best place because it is closer to the line.

Jeff gave an identical answer when the vector was placed at another location (see Fig. 6).

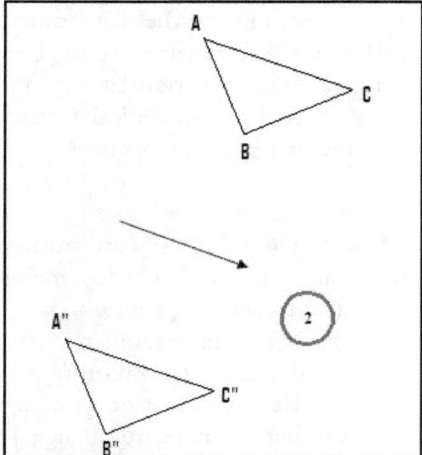

Fig. 6. Jeff's drawings of the images of triangle ABC.

Jeff's reasoning suggested that although he verbally stated that the vector determines where the translation was going to occur, he had a misconception about how he was supposed to use it. In both examples, he did not consider the direction and the magnitude of the vector; rather he used the distance between the pre-image point B and the vector. He then located the image at an approximated distance across the vector. However, it wasn't clear how he measured that distance. The distance was not taken perpendicular to the vector (see Fig. 5), and corresponding pre-image and image points were not equal distances from the vector (see Fig. 6). In both examples of translations Jeff preserved the orientation of the figures. Although it not completely clear how Jeff was thinking about the line (vector) in the case of translation, based on other data, we can say that Jeff's thinking in this instance was clearly different from his thinking about reflection. The first interview about reflection revealed that Jeff knew corresponding pre-image and image points would be equidistant from the line of reflection and that the segment connecting those points would be perpendicular to and bisected by the line of reflection. He also knew that reflection changes the orientation of the figures (that is if points ABC go clockwise their images A'B'C' will go counter-clockwise).

Jeff often referred to the vector as a line and did not seem to know the difference between a vector and a line. For a moment, he thought about the arrow of the vector, but he couldn't relate it to the direction of the translation.

At the end of the first clinical interview, several questions remained to be answered in the following teaching episodes. First, would Jeff conceive of translations as a one-to-one mapping of the plane onto itself rather than as a motion that is applied to a single object? Second, would Jeff be able to acquire a mapping understanding of translations and how would this understanding develop? And lastly, what critical factors would enable him to view translations from the mapping perspective?

4.2.2. *Translations as defined motions of a single object*

The analysis of the first set of clinical interviews formed a basis for the first teaching episodes. The findings showed that the major reason for Jeff to conceive of translation as undefined motion was his deficiency in understanding the parameter of the translation (the vector). The researcher inferred that it could be plausible for Jeff to modify his thinking by solving the tasks that emphasize the effect of translation rather than the translation itself. In this way, the researcher intended to change Jeff's attention from motion to *effect* of translation.

This section presents an analysis of the first teaching episodes used to assess Jeff's knowledge of translations, especially the knowledge of the parameter. Findings of these episodes suggested that developing an understanding about the concept of vector and its relationships to the pre-image and image points requires time and carefully designed experiences.

During the first teaching episode, Jeff explored translations using the *Geometer's Sketchpad*. Although *Sketchpad* does not draw arrows, it represents momentarily the vector as a moving set of dashes from the initial point to the end point when the vector is marked. During the teaching episodes the vector was shown on paper like a line segment with two endpoints and denoted by two letters, say vector AB. The researcher asked Jeff several questions to determine whether he knew the direction of vector or not. Jeff was able to tell the researcher that "A" would be tail of the vector and "B" would be the head of the vector. Jeff used his fingers to point out the direction of the vector.

Jeff was able to predict and perform translations accurately. Moreover, he was able to identify single and multiple vectors for translations. He knew that translations preserve the attributes of pre-image (e.g., size, shape, angle, angle measure, and orientation). Once Jeff became familiar with how the vector works, his attention turned on to the effect of translations. Performing composite translations and learning about the zero vector assisted Jeff's thinking about translations as the mapping of points on the plane. However, at the end of the first teaching experiment, Jeff still had a strong mental image of translation as motion.

4.2.2.1. Focus on the parameter and its relationships with the pre-image and image
The first teaching episode revealed that Jeff had difficulties in understanding the vector. Although he stated that the vector specifies the direction of a translation, he had to spend considerable amount of time to make a direct connection between the magnitude of the vector and the distance between the pre-image and image points. For example, he initially considered the distances between the end points of the vector and pre-image and image points rather than the length of the vector or distance between pre-image and image (see Fig. 7).

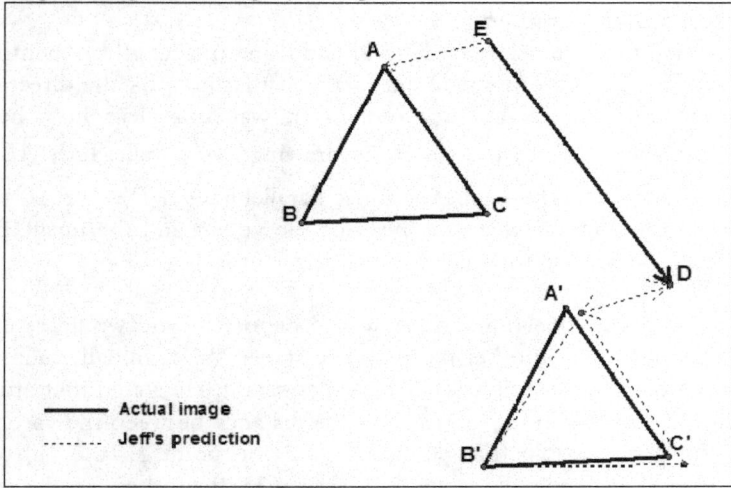

Fig. 7. Jeff's execution of a translation.

The following dialogue shows how Jeff decided where to place the image based on the given vector.

R: Could you please draw your prediction and explain to me how you determined where to locate the image triangle?

J: [He drew his prediction and said:] "I could do this [translation] actually pretty precisely if I wanted to." [He used the *point tool* and estimated the vertices of image triangle]. Then he said: "A would be here, B would be somewhere here and C would be here (see Fig. 7, the triangle with the dashed sides)." [Then he connected the vertices.] If I wanted to be more precise, I would draw a line between the points (A) to (E) and I would draw a parallel line that crosses the point (D). Then I would measure the distance (A) to (E) and put a point at the same distance from (D) to identify the location of (A'). I would also do the same thing for the other points and connect them to draw the image of triangle ABC. So all the lengths of sides, angles, and shape of the figure would be preserved.

By measuring the distance from E to A and using this length to identify the location of A' from the point $D,$ Jeff showed that he took the direction of the vector into consideration. However, it was not clear how he was considering the length of the vector by drawing two parallel lines $\overleftrightarrow{AE}$ (and $\overleftrightarrow{A'D}$). In other words, he might use those parallel lines $(\overleftrightarrow{AE}$ and $\overleftrightarrow{A'D})$ just to identify the distance between the head of the vector and the image points rather than considering the magnitude of the vector.

In response to the question "What would happen if you would point the head of the vector at another place on the screen?" Jeff initially mentioned what kinds of relationships would remain constant. He stated that after the dragging process, the sides, angles, and the distance between the vertices of the pre-image or whole triangle ΔABC and the point E (the tail of the vector) (see Fig. 7) would stay the same. On the other hand, he was thinking that the image of triangle would move relative to point D and everything about the image (e.g., size, angles, angle measures, orientation, and shape) would stay the same except its location. He said that it "depends on where I move it [the head of the vector], the image follows it. The pre-mage would not move." After that he actually dragged the head of the vector to validate his prediction. He also stated that the role of D (head of the vector) was to guide where the image would be located.

Jeff's explanation about the relationships between the head of vector and pre-image/image pairs suggested that he understood that the vector determines the direction of the translation and the translation preserves the attributes of the pre-image. However, Jeff was still considering the distance between pre-image/image points and the end points of the vector rather than the length of the vector.

Indeed, his response to the question "What would happen when the tail of the vector was placed at another point on the screen?" revealed that Jeff was considering the distance between the pre-image point and the tail of the vector ($|AE|$), and the image point and the head of the vector ($|A'D|$) rather than comparing the distance between A and A' ($|AA'|$) and the length of the vector ($|\overrightarrow{ED}|$). He couldn't predict accurately what would happen as a result of placing the tail of the vector at another point. He stated that the pre-image would move based on wherever the point E (the tail) moves to and the rest of them (the image and the head of the vector) would stay the same. Jeff actually dragged the tail of the vector and realized that the location of the image was changed rather than the position of pre-image. Jeff explained this experience in the following dialogue.

R: Could you please explain what happened?

J: Everything I thought was wrong. The pre-image and its relationship to E was static. That is, wherever E would move, that (pre-image) would follow. Just the same as this one (image). [Jeff dragged the tail of the vector again]. Wherever I move the point E, the relationships between the image and the point D act according to that. If I move E closer to A, D gets closer to A'.

Jeff's explanation suggested that he was initially thinking there was a fixed relationship between the end points of the vector and vertices of both pre-image and image (e.g., distances $|EA|$, $|DA'|$, $|EB|$, $|EC|$, $|DB'|$, $|DC'|$). Since he thought those distances were fixed, he expected the same relationships would be maintained after he dragged the tail of the vector. However, he did not consider what would change about the vector (e.g., its magnitude or direction) as a result of changing the location of the tail and how this would affect the translation eventually.

In the case of identifying the *effect* of placing the entire vector to another location without making any changes on the magnitude or the direction of the vector, Jeff initially predicted that the pre-image and image would move simultaneously. This also suggested that Jeff was still considering the relationship between the pre-image/image points and the end points of the vector as static. However, he then changed his mind and stated that only the *image* would move and *pre-image* would stay where it was located. Yet, he wasn't thinking about what changes would occur when the entire vector was located at another place and what that means. Although it was mentioned in the question that there would be no change in the direction or the magnitude of the vector, Jeff did not take them into consideration. Jeff might think that if the vector moves so does the image since the vector specifies where the image would be placed.

After Jeff located the entire vector at another place, he saw that nothing changed. He said, "Oh! The vector is almost independent of everything. It is not depending on triangles at all." This experience was critical in terms of understanding the role of the vector. Once Jeff's attention moved onto the effect of the action he took, he then concentrated on what had changed and what remained the same by locating the vector at another place. He stated that he changed the location of the vector and that nothing moved. Observation of the effect of changing the vector helped Jeff begin to clearly see the relationship between pre-image/image pairs and the vector. He understood that even though the location of the image was determined by the vector, the position of the vector was independent of both pre- image and image and did not affect the pre-image or image in any way. However, Jeff's reasoning suggested that he was still conceptualizing a motion-oriented since he expected a movement as a result of his action.

Focusing on the *effect* of changing the magnitude and the direction of the vector facilitated Jeff's understanding of the relationship between pre-image/image pairs and the vector. Jeff stated that if he decreased the length of the vector, the image would act corresponding to that and get closer to the pre-image. He also added that the length of the vector ($|\overrightarrow{ED}|$) and the distance between A and A' ($|AA'|$) were equal (see Fig. 7). He dragged the vector and showed that they were the same. This was a breakthrough in Jeff's thinking about translation since he was able to understand that the distance between the pre-image and image points was defined by the magnitude and direction of the vector.

Further explorations with the Geometer's Sketchpad helped Jeff begin to understand the *effect* of changing the vector (e.g., magnitude, direction and location) on both pre-image and image. Earlier he was thinking that changing the location of the entire vector without changing its magnitude and direction would affect where the image would be placed. However, his experience with the Geometer's Sketchpad helped him realize that the location of the vector has no effect on pre-image and image. Furthermore, he was able to understand that the length and the direction are key aspects of a vector that define a translation.

4.2.2.2. Focus on the effect of translation

The *effect* of multiple translations helped Jeff develop a better understanding of the concept of vector in relation to translations. For a given task (see Fig. 8), Jeff stated that there would be infinite number of vectors and the only thing that matters was the length of the vector. For a single translation, Jeff gave several examples, such as $\overrightarrow{IS}$, $\overrightarrow{FO}$, $\overrightarrow{GP}$ and said, "million other points that can be used to correspond to the same length." He also stated that the effect of translation under the vectors $\overrightarrow{IS}$ and $\overrightarrow{SI}$ would be different since the location of the image would not be the same in these two cases. Jeff's explanations also suggested that he understood that to have the same effect on the pre-image, one needs to consider the length and the direction of the vector.

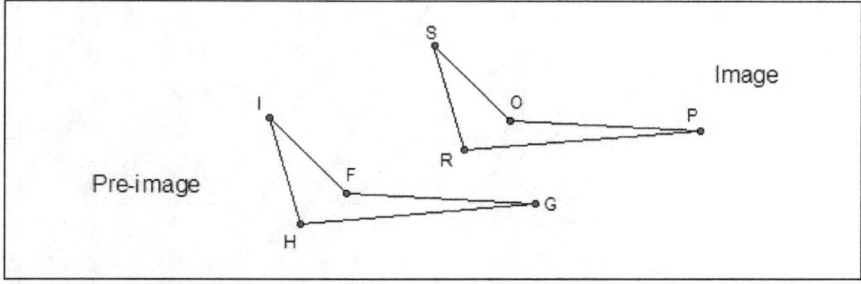

Fig. 8. Identifying vectors (Adapted from CAS-Intensive Mathematics Project (2004) materials).

Jeff was able to identify multiple vectors that would produce the same effect as a single vector. He stated that what he needed to do was to find the length of a single translation vector and use enough other vectors to equal the other line [the vector]. Initially, he drew a single translation vector and added up several vectors that were parallel to the previous one (see Fig. 9). He further constructed multiple vectors, which were not parallel to the original vector. Jeff seemed to consider the length of the vector. However, how he located the starting point of the first vector and the ending point of the last vector showed that he also considered the direction. Once he provided the second example, he went back to his initial example and said that it did not have to be parallel to the original vector.

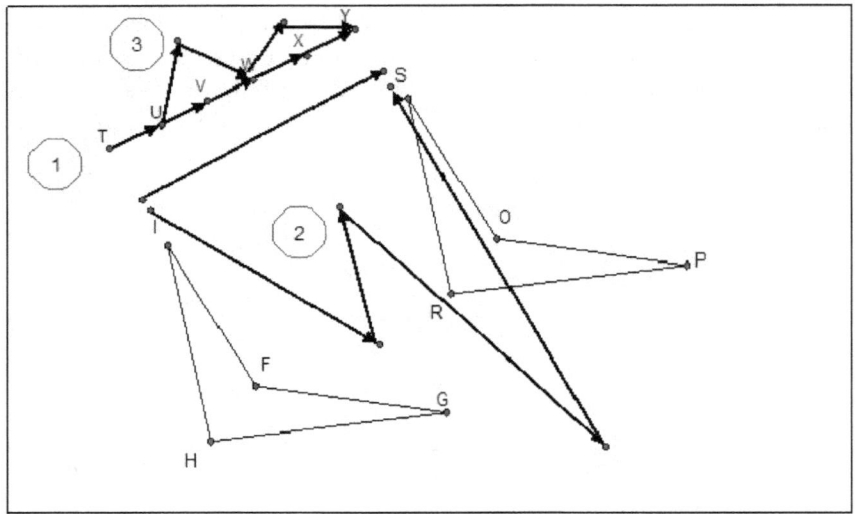

Fig. 9. Jeff's examples of multiple translations.

The *effect* idea helped Jeff see that the final location of the image points in the plane was critical for identifying the result of translations. This was a crucial moment for Jeff since translation itself became a procedure and his attention was no longer on the action of translation but on the result. For example, on the previous task (see Fig. 9) Jeff was able to realize that by using a composition of different vectors, the pre-image would follow a different path, but it might end up at the same location. Jeff stated that the location of the image was critical in terms of having the same effect on the pre-image. Consideration of "location" was also a vital component of understanding transformations as mapping. Jeff noticed that as long as he knew the location of the image, he could draw infinite number of vectors having the same effect on the pre-image.

Although Jeff's attention was on the *effect* of translations, he was still conceiving a translation as a motion of a single object defined by the vector. For example, in response to the question "whether it was possible to map the pre-image (polygon IHGF) onto itself" Jeff initially tried to move the pre-image polygon (IHGF) onto its image polygon (SRPO) by decreasing the length of the vector (see Fig. 9). However, he said: "you cannot move the pre-image by using the vector." His answer indicated that he was still concentrating on the motion aspect of the translation. Then he decreased the length of the vector until he got a zero vector and realized that it would be possible to map the pre-image onto itself. However, he was reluctant to accept a zero vector when he was asked whether it could be possible. He was thinking that a vector has two points and if the magnitude of the vector was zero, there wouldn't be a vector at all, so the pre-image would not move. Jeff also wasn't sure about whether there would be a translation when the vector was zero. He initially thought that there wouldn't be a translation, but then he changed his mind and said:

> "Perhaps there is but it just stays right where it is. Perhaps, if the vector was zero and I translate this image maybe I would get all these next images underneath it. More and more images would be created underneath it."

Jeff's confusion about the zero vector was a turning point since it was contrary to his previous beliefs. That is, if there was a zero vector, translation wouldn't be explained through motion anymore. There should be another explanation.

At the end of the first teaching episodes, the questions that need to be answered for the following teaching experiments for Jeff was still the same. That is, whether Jeff would conceive translations as mapping of all points in the plane rather than as a motion that is applied to a single object, and if he would, how he would proceed and what are the critical factors that would affect his progress.

4.2.3. Translations as a defined motion of all points on the plane

The first teaching episode had indicated that Jeff had a motion conception of translations. However, his conception was not the same as in the initial interviews. Once he became familiar with the parameter, he began to think that translations were defined by the vector. He was considering translations as a motion that is applied to a single object. The next teaching episodes focused on the domain of the translations as all points in the plane. Prior occasions had revealed that Jeff considered the domain of the translations as a single object, which was independent of the plane.

4.2.3.1. Understanding the domain of the translations

The domain as all points in the plane was new for Jeff. Jeff had various difficulties trying to conceptualize the domain as all points in the plane. This section presents an analysis of the second teaching episode dedicated to assessing Jeff's understanding of the domain of the translations.

At the beginning of the episode, it appeared that Jeff seemed to consider only the visible points on the screen for a translation. For example, when a problem asked to translate a triangle with a yellow interior, Jeff considered selecting the vertices and sides of the triangle only, instead of selecting the whole triangle (see Fig. 10). He also stated that there would be no difference between the two selections. After the translation, Jeff realized that he did not translate the triangle interior. The color of the interior area helped him see that he needed to consider those points as well. Jeff might see the computer screen like an empty background, rather than consisting of an infinite number of points. Since Jeff was thinking the triangle as not part of the plane itself, that is independent of the plane, he was selecting the triangle only and moving it on the plane rather than selecting all points in the plane. During the first teaching episode, the researcher had asked him whether there was a difference between moving a single object on a paper and moving the whole paper. At that time Jeff said that moving the whole paper translates everything on the paper. He further stated that it was like highlighting everything on the screen and then translating them. In this teaching episode, he was still reluctant to consider all points in the plane when translating.

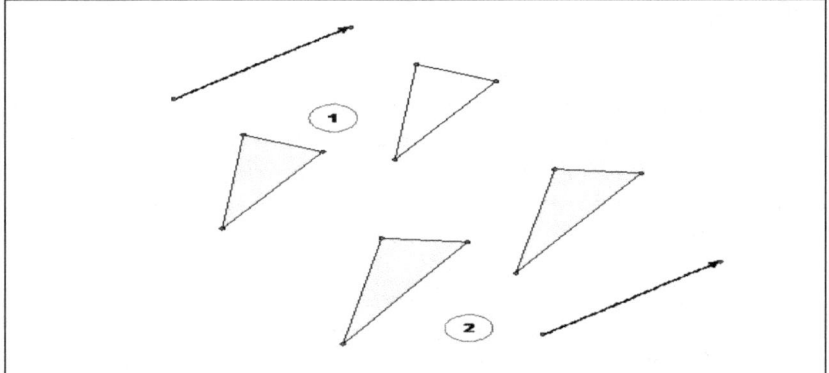

Fig. 10. Jeff's examples of translations.

It became apparent that Jeff was considering the domain as a single object under a translation when a translation task included two triangles. Jeff stated that he would only translate one of the triangles since the problem asked him to do so. He added that the other triangle has no connection to the yellow triangle. His reasoning suggested that he was thinking of the triangles individually (laying on the plane) independent from the plane and from each other rather than being part of the plane.

Jeff's prior experience was also another factor preventing him from accepting that a translation applies all points in the plane simultaneously. Although he accepted that mathematically it might be accurate to say that a translation applies to all points in the plane, it was still contrary to what Jeff believed was true. For example, he said:

> "This is what I believe translation: We translate ourselves all day long. I move from here to down cross the street to my house. I translate myself. I am translating upon composed transformation; reflection, not reflection, maybe rotation, I translate and I rotate. That is how I get one place to another. So by me translating, not

> everything in the world moves; just me translated. If you move everything in the plane, actually nothing moves."

Once again, Jeff's reasoning showed that he was using non-mathematical clues. His real life experience made him think that translation was a movement and objects were independent from each other and the plane.

Jeff, while doing an activity focused on the understanding of geometric figures as part of the plane and the domain as all points in the plane, began to conceive of what translating all points in the plane to other points in the plane really is. Initially, he believed that when the whole plane moves, the coordinate system would also move with the plane so there would be no change in the location of geometrical objects. However, when he realized that the coordinates are independent of the plane, he was able to see that when the geometric figures were translated to other points, their location on the plane changed accordingly. This was a crucial experience for Jeff in understanding that all points in the plane were mapped to other points in the plane.

4.3. Jeff's current understanding of translations

Towards the end of the study, Jeff became more confident and seemed to understand translations. He was able to recognize, describe, and perform single and multiple translations on geometric figures. More importantly, he was thinking that translations applied to all points in the plane and that the vector defines the translation.

This section presents an analysis of the last clinical interview dedicated to assessing Jeff's current reasoning patterns of rigid translations without attempting to change them.

4.3.1. Recognition and description

Jeff was able to recognize whether two given images represented a translation or not. For example, when he was asked to identify whether two given squares (see Fig. 11) can be an example of a translation, he said that it could be if the corresponding points are in the same order with the pre-image (e.g., ABCD and A'B'C'D'). He knew that rigid translations preserved the features of the geometric figures and saw that all sides, angles, orientation, and the shape of the image had been preserved. He was able to identify both single and multiple vectors for this translation and specified the direction and the length of the vector. Jeff also mentioned that this transformation cannot be a rotation at the same time since the corresponding points would not be the same.

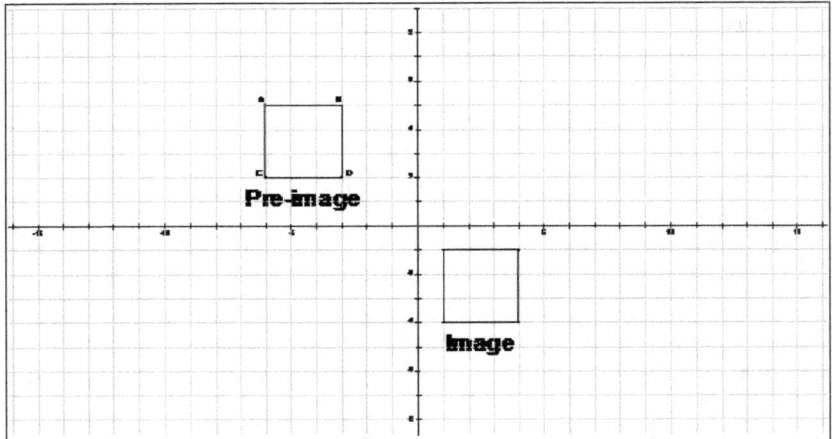

Fig. 11. Can it be an example of a translation?

4.3.2. Execution and representation of translations

When Jeff was asked to perform a translation including three geometric figures, he accurately executed the translation using the vector. When he was asked if he would consider translating all points in the plane, he indicated that all the figures would be translated and all the relationships would be preserved under the translation, such as the distance among the figures, size, angle, and orientation of the shapes. Although Jeff did not translate the vector itself, when he was asked if he would translate it, he said he would since he was supposed to translate all points in the plane. However, he mentioned that initially he considered the vector as an abstract guide rather than a physical object which is part of the plane.

Jeff seemed to understand the effect of changing the vector on the pre-image and image. He explained that the location of the vector is independent of everything, which means that it affects neither the pre-image or image in any way. He indicated that direction and the magnitude are features of a vector, so if they are changed, the location of the image would change. He gave an example and stated that if he decreased the length of the vector, the images of trapezoid, triangle and parallelogram would get closer to their pre-images.

Jeff was able to predict the inverse of the translation without actually having to perform it. When he was given an image and a vector and asked to find the pre-image, he used the opposite direction length of the vector and identified the location of the pre-image. Jeff had tried to solve a similar problem in the first clinical interview, but he couldn't find the answers at that time.

Jeff was able to reason through the composition of two or more transformations and determine the properties that would be preserved. He was able to represent his thinking using various function notations. For example, when he was asked to represent a composite translation with a function notation using multiple and single vector, he wrote:

$T_{\overline{AB}}(KLMN)$ $T_{\overline{CD}}(K'L'M'N')$ $T_{\overline{EF}}K''L''M''N''$ $K'''L'''M'''N'''$

$T_{\overline{SO}}(KLMN)$ $K'L'M'N'$

Regarding the first representation, Jeff explained that the trapezoid (KLMN) was first translated based on the vector $(\overrightarrow{AB})$ and then the new trapezoid (K'L'M'N') was translated based on the vector $(\overrightarrow{CD})$ and finally (K''L''M''N'') translated under the vector $(\overrightarrow{EF})$. When he was asked to identify a single vector that has the same effect on the pre-image, Jeff drew the vector $(\overrightarrow{SO})$ and represented a single translation by the notation given above.

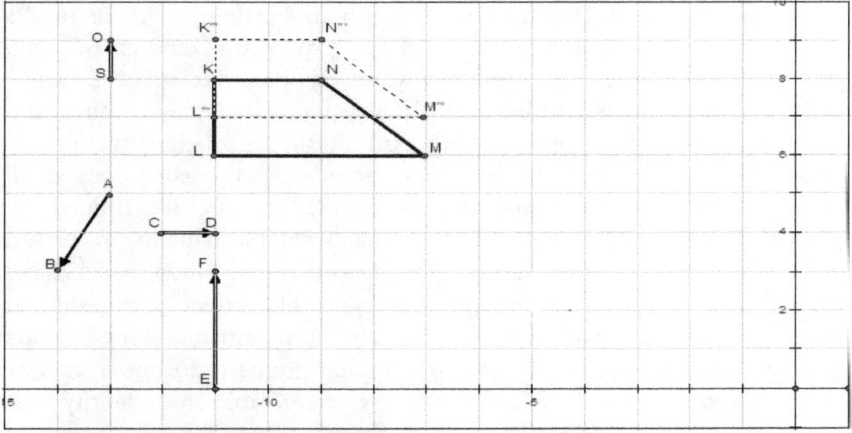

Fig. 12. Jeff's examples of composite translations.

Towards the end of the study, Jeff was able to predict the effect of a translation without actually performing it. He was able to identify single and multiple vectors that produced the same effect on a pre-image. He developed a strong understanding of the *parameter* and the *effects* of changing the *parameter* on the translation.

Furthermore, he was able to understand a composition of translations without actually performing them. He was able to represent and understand composite transformations by using function notation. He was able to understand the properties of translations in general without focusing on a specific translation situation. Although Jeff began to think that translations apply to all points in the plane, he was unable to conceive of geometric figures as part of the plane. Consequently, he continued to operate from the motion perspective of translations. This was one of the main difficulties that prevented Jeff from thinking of translations as a mapping of all points in the plane to other points in the plane.

5. Discussion

The broader study of which Jeff's study was a part investigated four pre-service elementary teachers' understanding of rigid transformations before, during and after their experience within a computer-based learning environment and how their thinking evolved towards seeing transformations as mappings of the plane onto itself.

The findings of these teaching experiments suggested that the four prospective elementary teachers involved in this study predominantly had a motion understanding of transformations which was based on a movement of an object from one point to another point. Furthermore, the results revealed that the participants' growth in conceptualizations of rigid geometric transformations as motion towards mapping depend on a variety of factors and the transition between the two kinds of understandings is a complex process which requires a long term careful designed instruction.

The same as in the case of Jeff, understanding parameters helped all participants realize that transformations were actually determined by parameters. Focusing on the parameters allowed participants to reason about the effect of transformations. By the end of the study, participants understood how changing the parameters would affect the result of transformations. They were considering that the parameters would define the transformations and were used to mapping points onto new locations under a transformation. Moreover, they were able to identify the parameter(s) for given pre-image and image using the properties of transformations. However, they still considered transformations as motions of individual objects though they were defined by the parameters.

The findings revealed that understanding domain as all points in the plane was a new and a difficult concept for the participants. Initially all the participants were thinking the domain as a single geometrical object (or a point or a line) rather than all points in the plane. Although they considered the relative distances and positions of the points in a single object as a result of the transformations, they did not take them into consideration for the multiple discrete geometric figures. By the end of the study Jeff was considering the domain as all points on the plane for translations and rotations only. He was thinking that these transformations would preserve the relative distance and position of the points both within the figures and among the figures. However, he was reluctant to think about the domain as all points in the plane for a reflection. This suggested that the concept of domain was still in the progression for Jeff though he developed a significant amount of understanding throughout the study.

5.1. Motion versus Mapping Understanding of Transformations

All participants conceived of transformations as motion. They explained that a translation, reflection and a rotation could be described as a movement of all points in the plane rather than mapping of the plane onto itself. Because points were viewed, in this case, as separate from the plane, participants' limited understanding of the plane interfered with an interpretation of transformations as mapping. For example, participants had difficulties in understanding the meaning of mapping. Since they were considering the result of a transformation as a motion of an object or points, there would not be a translation, reflection or a rotation without movement. The motion expectation directed participants' attention and was an important aspect of their definition and mental image of transformations.

An important step in the development of their understanding of transformation was the realization of the role of the parameters that determine the different transformations. Once participants became familiar with the parameters, they began to show some structured aspects of understanding of the transformations. They were thinking that every point has one corresponding point which was determined by the parameters. However, their understanding of plane in relation to geometric figures was still the main difficulty for them to understand the nature of mapping of the plane. Since they were thinking that geometrical objects were not part of the plane, the mapping was like placing the objects in a different location on the plane.

5.2. The mapping process vis-à-vis understanding the plane

The results of the teaching experiment revealed that participants conceived of the plane as a background on which geometric figures are placed. In other words, they were considering the geometric figures separate from the plane in which they were located. This is different from the perspective of contemporary mathematics in which points (or lines or figures) are located at particular places on the plane but they are not separate from places where they are located and they are actually subsets of the plane (Smart, 1998; Edwards, 2003). Since participants were thinking of objects as individual entities, which were independent from the plane and from each other, they applied the transformations only to discrete objects in the first two stages (see Fig. 2). Therefore, when they were provided multiple discrete geometric figures to be translated, they did not consider all the initial geometric relationships (e.g., relative distances and angles) among the figures.

Possible reasons became apparent. For example, Jeff's efforts to make sense of mathematical situations through using his experience in the physical world seemed to be one of the sources of his conception. He used an example from four-dimensional world (3-dimensional space and time) and applied it to the two dimensional plane. For example, he explained translations with an analogy to how people transport themselves from one place to another. In this case, he considered people as individuals completely independent from one another. This partially explained why Jeff was thinking that geometric figures sit on the plane rather than being subsets of the plane.

Jeff's past experience in mathematics classes seemed to be another reason why he thought that geometric figures are not part of the plane. He mentioned that in his past experiences, the examples, especially the ones that included concrete materials, were presented individually on flat surfaces. Thus, the relationships between the object and the surface were not emphasized explicitly. It is possible that because of this Jeff considered the figure and the plane as independent from each other. This was one of the major difficulties in Jeff's understanding of the nature of mapping of the plane.

By the end of the study, although Jeff was thinking that the plane was static and consisted of infinite number of distinct points, he held the belief that everything (e.g., geometrical objects) was independent of the plane except for the coordinate system. According to Jeff, when a transformation of any type is performed, the objects themselves are mapped onto new coordinates of the same plane. Understanding the geometric figures as subsets of the plane, however, with certain relationships among points, seemed a difficult concept for Jeff throughout the study.

5.3. The theoretical model

Based on the results of the teaching experiment, a theoretical model was developed to describe the participants' growth in understanding rigid transformations (see Fig. 2). The model displays the transitional understandings two of the four prospective teachers constructed as they progressed from a motion understanding toward a mapping conceptualization of rigid transformations and shows some connections among these two understandings. The other two teachers did not advance as much in their development and in their case there was only evidence for the initial part of the model, that is, moving towards the notion of defined motion of a single object.

We can use elements of schema theory (Skemp, 1987; Marshall, 1993) to understand the different nature of the difficulties Jeff experienced moving on one hand from the initial level of undefined motion of single objects to that of defined motion of single objects, to moving to the conception of defined motion of all points in the plane on the other hand. According to Marshall (1993), "The schema is the organizing structure that draws similar examples, features, plans, and procedures together and binds them into a cohesive unit." (p. 266). Skemp (1987) points out that a schema "integrates existing knowledge, it acts as a tool for future learning and it makes possible understanding" (p. 24). We can interpret the progress of understanding of Jeff going from the first level to the next as an example of assimilating his idea of translation into a schema of transformations as motion, while at the same time developing better internal organization of the schema to see transformations as defined motions of objects, where the transformations are defined by their corresponding parameters. As Skemp (1987, p. 29) points out, both assimilation into a schema and developing a better organization of the schema contribute to better understanding.

The transition from the level of defined motion of an object to that of defined motions of all points in the plane is of a different nature in that it requires thinking about transformations as mappings of the whole plane, rather than motion of objects, that is, it requires a change of schema. Because schemas are such major instruments of adaptability, and provide such effective organization of existing knowledge, there is a strong tendency towards the perpetuation of existing schemas (Skemp, 1987, p 27). We can interpret Jeff's difficulties in going from the second level to the third as an example of the intrinsic difficulties involved in the reconstruction of a schema.

Although this model might not be applicable for all prospective teachers, it is still a useful tool for understanding how participants perceived and made meaning of rigid transformations based on the tasks that were asked during the teaching experiment. It was also an effective model for understanding how participants' understanding was developed and the critical factors that facilitated their progression from one level to the next level of understanding.

6. Implications for Future Research

Although this model became apparent for two of the participants at the end of the study, it was still unclear whether it would completely fit the other two prospective teachers that participated in this study since they were not at the same level with Jeff and the other participant. Our conjecture at this point would be positive, if they had enough time and experience to study geometric transformation. Both of these participants began to conceive of transformations as undefined motion and then progressed through understanding them as defined motion of a single object. However, at the end of the study, they still had a motion conception of transformation and their conceptions of domain and plane were incomplete. Future research needs to be conducted to identify what other concepts are involved in conceiving transformations as mapping. Furthermore, what mathematical understanding does one need to develop to be able to internalize those ideas and how does this happen? Also, the model created as a result of this teaching experiment could be examined again to understand whether or not this framework works for other cases and in what ways it could be revised or improved.

References

CAS-Intensive Mathematics Project. (2004). *Technology-Intensive Secondary School Mathematics Curriculum* (CAS-IM, NSF Grant No. TPE 96-1802). Retrieved April 19, 2009 from http://www.ed.psu.edu/casim/

Desmond, N. S. (1997). The geometric content knowledge of prospective elementary teachers. (Doctoral Dissertation, University of Minnesota, 1997) *Dissertation Abstracts International, 58*(08), 3050A.

Edwards, L. (2003, February). The nature of mathematics as viewed from cognitive science. Paper presented at 3rd Congress of the European Society for Research in Mathematics, Bellaria, Italy.

Edwards, L., & Zazkis, R. (1993). Transformation geometry: Naïve ideas and formal embodiments. *Journal of Computers in Mathematics and Science Teaching, 12*(2), 121-145.

Engelhardt, P. V., Corpuz, E., Ozimek, D., & Rebello, N. S. (2003). The teaching experiment- What it is and what it isn't. Paper presented at Physics Education Research Conference, Madison, WI, U.S.A.

Flanagan, K. (2001). *High school students' understandings of geometric transformations in the context of a technological environment.* Unpublished doctoral dissertation, Pennsylvania State University.

Harper, S. R. (2002). Enhancing elementary pre-service teachers' knowledge of geometric transformations. (Doctoral Dissertation, University of Virginia, 2002). *Dissertation Abstracts International:* AAI3030678.

Hollebrands, K. (2003). High school students' understandings of geometric transformations in the context of a technological environment. *Journal of Mathematical Behavior, 22*(1), 55-72.

Law, C. K. (1991). A genetic decomposition of geometric transformations (Doctoral Dissertation, Purdue University, 1991). *Dissertation Abstracts International, 52*(06), 2057A.

Leinhardt, G., & Smith, D. (1985). Expertise in mathematics instruction: Subject matter knowledge. *Journal of Educational Psychology, 77*(3), 247-271.

Ma, L. (1999). *Knowing and teaching elementary mathematics: Teachers' understanding of fundamental mathematics in China and the United States.* Mahwah, NJ: Lawrence Erlbaum Associates.

Marshall, S.P. (1993). Assessment of rational number understanding: A schema-based approach, in T. P. Carpenter, E. Fennema, T. A. Romberg (Eds.), *Rational Numbers: An Integration of Research*, p. 261-288. Hillsdale, NJ: Lawrence Erlbaum Associates.

Mayberry, J. (1983). The van Hiele levels of geometric thought in undergraduate preservice teachers. *Journal for Research in Mathematics Education, 14*(1), 58-69.

National Council of Teachers of Mathematics. (1989). *Curriculum and evaluation standards for school mathematics.* Reston, VA: NCTM.

National Council of Teachers of Mathematics (2000). *Principles and Standards for School Mathematics.* Reston, VA: NCTM.

Parsons, R. R. (1993). *Teacher beliefs and content knowledge: influences on lesson crafting of pre-service teachers during geometry instruction.* Unpublished doctoral dissertation, Washington State University.

Simon, M. (1995). Reconstructing mathematics pedagogy from a constructivist perspective. *Journal for Research in Mathematics Education, 26*(2), 114-145.

Skemp, R. R. (1987). *Psychology of learning mathematics.* Hillsdale, NJ: Erlbaum.

Smart, J. R. (1998). *Modern geometries.* Pacific Grove, CA: Brooks/Cole Publishing Co.

Steffe, L.P., Thompson, P.W.(2000). Teaching experiment methodology: Underlying principles and essential elements. In A. E. Kelly & R. Lesh (Eds.), *Handbook of research design in mathematics and science education* (pp. 267-306). Mahwah, NJ: Lawrence Erlbaum Associates.

Yanik, H. B. (2006). *Prospective elementary teachers' growth in knowledge and understanding of rigid geometric transformations.* Unpublished doctoral dissertation, Arizona State University.

21 THE OPPORTUNITY GAP[21]

There are considerable differences of performance in national and state tests among different ethnic groups, between students from low income families and high income, and among students whose parents have different levels of schooling. Often the unequal performance of Latino and African American students compared to European American students is described as an achievement gap. It is not uncommon to use statements like the following to describe the situation: "Students of color continue to lag behind white students and some Asian students, and the so-called academic achievement gap still exists" (a state superintendent of public instruction, as quoted by Heffter, 2006). "Across the U.S., a gap in academic achievement persists between minority and disadvantaged students and their white counterparts" (National Governors' Association, 2005). What kind of images do we form about the students who lag behind after reading such statements? What kind of assumptions, conscious or subconscious, do we make about their capacity of learning? Do we even ask why is their performance lower? What unquestioned assumptions do we make about the educational system?

Blanket statements about the low performance of certain groups of students in our schools that do not mention the underlying causes may reinforce prejudices and stereotypical images some people have about such groups. Unfortunately, such prejudices are not uncommon. Some authors even claim that Latino and African American students are "less teachable."

---

[21] Flores, A. (2008). The opportunity gap. In R. Kitchen and E. A. Silver (Eds.), *Promoting high participation and success in mathematics by Hispanic students: Examining opportunities and probing promising practices* (p. 1-18). Washington, DC: National Education Association. Reprinted by permission of TODOS: Mathematics for all.

For example, Greene and Foster (2004) state that being minority is a disadvantage students bring to school. They claim that as the percentage of White (non-Hispanic) students decreases in a school, the "teachability index" decreases too.

Finding a proper way to frame a problem often can give us not only a better understanding of it, but also show us better ways to address the problem and solve it. In education it is important not just to address the symptoms, but to understand, make explicit, and address the underlying causes as well. Looking at school and district practices and policies that cause to a great extent the disparity of performance in mathematics among different groups of students in schools can help us understand how to provide equal learning opportunities for all students.

In this chapter, I will first present some data that show striking and persistent differences of performance in tests among different groups of students in schools. I will then present evidence that shows that students from some groups are not as likely to have the same opportunities to learn mathematics in our schools as students from other groups.

The achievement gaps

In this section, I will present data that show gaps in mathematics learning that are important in size. These gaps have been quite persistent over the last three decades. The gap in test performance in mathematics is a wide one. At eighth-grade, 91% of African-American, 87% of Latino, and 85% of Native American students are not proficient in math, as measured by the National Assessment of Educational Progress (NAEP). This stands in stark contrast to the performance of European American and Asian American students (see Figure 1). Also, as Figure 2 shows, 12th grade Latino and African-American students perform on NAEP's math assessment at the same level as 8th grade White students (Wilkins & Education Trust staff 2006). The disparity of performance is greater on short constructed-response items than for multiple-choice items, and for extended constructed-response tasks the difference is even bigger. For example, in the 2000 mathematics NAEP test the percentage of correct answers given by African American 8th-grade students in multiple choice questions was 72% compared to that of correct answers given by European American students. For short constructed items, it was 63%, and for extended constructed-response tasks it was only 32% (Strutchens, Lubienski, McGraw, & Westbrook, 2004, p. 279). That is, the more complex the kind of answer required, the bigger the difference in achievement.

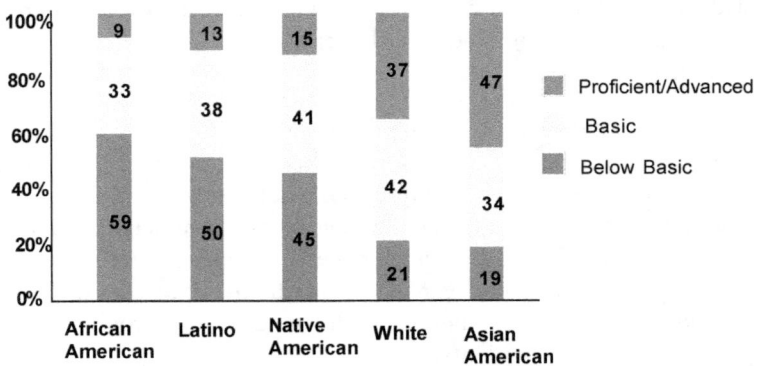

Figure 1. 2005 NAEP Grade 8 by ethnicity
Source: National Center for Education Statistics, NAEP Data Explorer as cited by Haycock, 2006, Slide 17

The gap has also been very slow to close. There was some narrowing of the gap in the 1970s and 1980s, but since 1988 the gap widened somewhat or remains about the same (see Figures 3 and 4). There is also a considerable gap in test performance between students who come from poor families and those who come from non-poor families (Figure 5). There are also clear and persistent differences in the performance of students when grouped according to the education level of their parents (Figure 6). Differences in parents' school achievement tend to be reproduced in the schools, rather than erased.

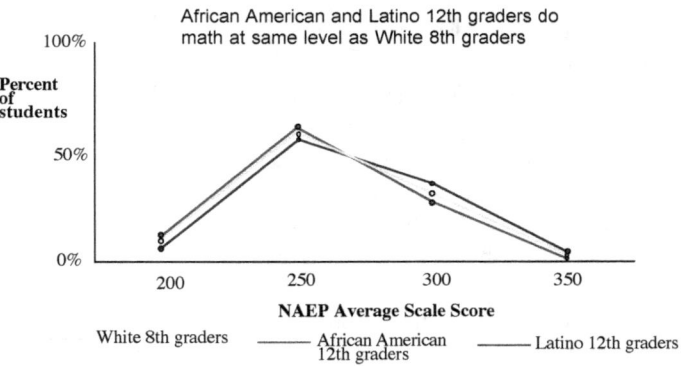

Figure 2. NAEP performance for different groups.
Source: Wilkins et al., 2006. ©The Education Trust. Used with permission.

244

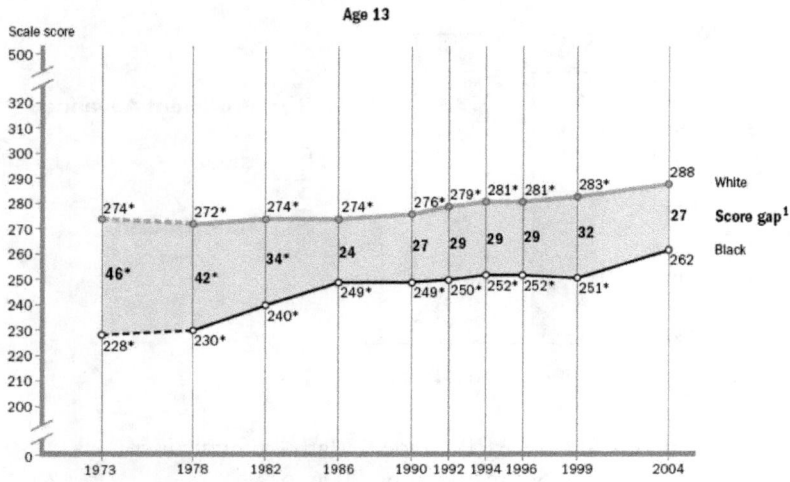

Figure 3. Trends in average mathematics scale scores and score gaps for White students and African American students age 13, 1973–2004
Source: Perie, Moran, & Lutkus, 2005, p. 42

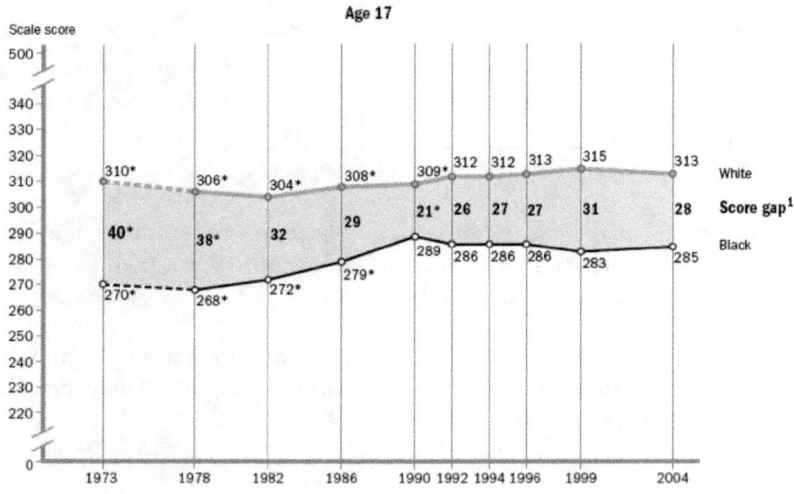

Figure 4. Trends in average mathematics scale scores and score gaps for White students and African American students age 17, 1973–2004
Source: Perie, Moran, & Lutkus, 2005, p. 42

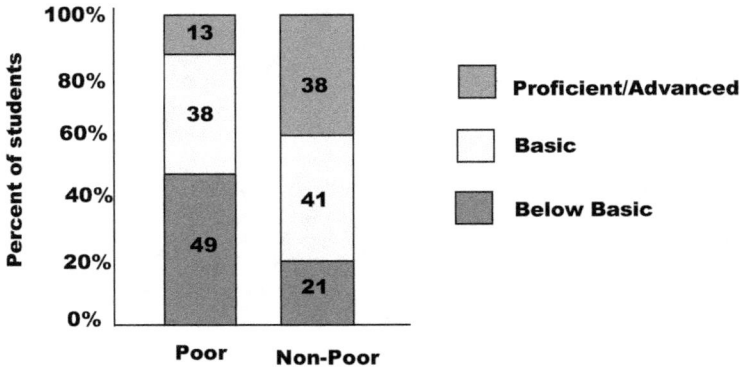

Figure 5. 2005 NAEP Grade 8 Math by family income
Source: Wiener, 2006, p. 3.

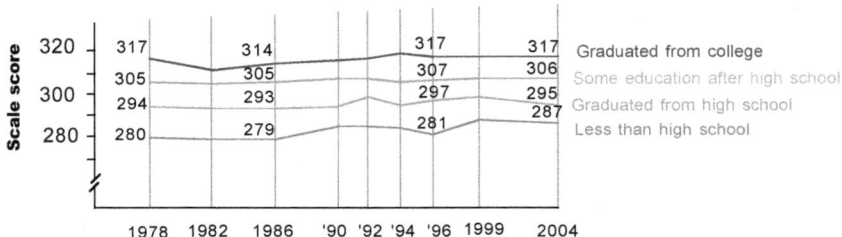

Figure 6. Trends in average mathematics scale scores for students age17, by student-reported parents' highest level of education: 1978–2004
Source: Perie, Moran, & Lutkus, 2005, p. 45.

It is thus clear that students from some groups are not learning as much in our schools as students from other groups. The important questions are *Why do such disparities exist? What are the causes?*

The opportunity gaps

A very different way to describe the distinct experiences among students in schools is to understand how Latino and African American students are less likely than White students to have qualified teachers, high quality mathematics instruction, and appropriate use of resources. For example, African American and Latino students are less likely than White students to

- Have teachers who emphasize reasoning and nonroutine problem solving,
- Have access to computers, and
- Have teachers who use computers for simulations and applications (Strutchens & Silver 2000)

African American and Latino students often experience a lesser form of education, in mathematics and in general (Diversity in Mathematics Education Center for Learning and Teaching, 2007). In contrast, White students usually have adequate mathematics course offerings, qualified mathematics instructors, quality mathematics curricula, and teachers who respect their culture and hold high expectations of them.

The National Council of Teachers of Mathematics (2005) answers the question of "How can we close the achievement gap in mathematics" by stating that all students "should have equitable and optimal opportunities to learn mathematics free from bias," and that "all students need the opportunity to learn challenging mathematics from a well-qualified teacher who will make connections to the background, needs, and cultures of all learners." The solution to the achievement gap is thus framed as opportunity to learn.

In the following pages, I will describe ways in which many Latino and African American students are less likely to have equal opportunities to learn. I will illustrate how many low-income and minority students are systematically shortchanged by the educational system by having less access to experienced and qualified teachers, by facing low expectations, and by receiving less funding per student. These three dimensions of opportunity are of course not the only important factors. Classroom practices, for example, have been empirically demonstrated to be linked to student performance (Hiebert and Grouws, 2007). Other important factors are briefly mentioned in the final comments section. For each of the three factors discussed in this chapter, I provide a brief rationale of why that factor is important at the beginning of the corresponding section.

Gaps in opportunities to have qualified and experienced teachers

Qualified teachers who are committed to the learning of their students are the single most important factor for students' success (Darling-Hammond, 2001). As can be documented by multiple research studies, a good teacher can make a big difference (Darling-Hammond, 2000). Having several experienced and highly qualified teachers in a row can improve dramatically the performance of students (Sanders and Rivers, 1996). By pointing out inequities in the distribution of teacher quality among schools, my intention is not to denigrate the many outstanding, talented and dedicated teachers who are teaching our most disadvantaged children, often under very hard conditions. Quite the contrary, the intention is to underline the importance for all children to have access to their fair share of such exemplary teachers. By stressing the importance of experienced and qualified teachers, the intention is not to put the blame on inexperienced teachers, but to focus on structural inequities in opportunities for students from different groups at the school, district, and state level. In this section, I will describe how African American and Latino students and low income students are shortchanged in their opportunities to have access to teachers who have experience, to teachers who are well qualified in mathematics, and teachers who are generally well prepared.

Access to experienced teachers

African American and Latino students are less likely to have experienced teachers than their European American counterparts (Mayer, Mullens, & Moore, 2000). Classes at schools that serve mostly African American and Latino students are twice as likely to be taught by teachers with three years of experience or less than classes at schools where there is a majority of White students (Figure 7). Similarly, classes at high poverty schools are more likely to have inexperienced teachers than at low poverty schools (Figure 8) (Mayer, Mullens, & Moore, 2000). In schools that are hard to staff and with high turnovers of teachers, students do not always have a permanent teacher. For example, at the Brooks Academy in Phoenix, where more than 99% of the students are Hispanic, Native American, or African American, and where 99.5% of the students qualify for free or reduced price lunch, one fourth-grade class had 17 substitute teachers before a full time teacher was hired in January (Kossan, 2006).

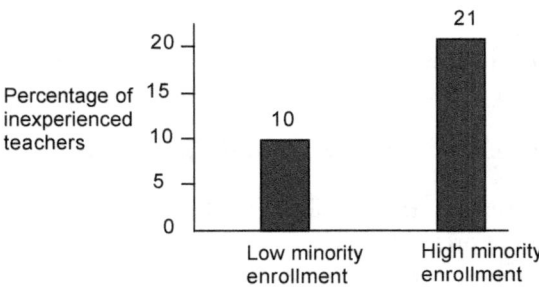

Figure 7. Percentage of inexperienced teachers by minority enrollment.
Source: Mayer, Mullens, & Moore, 2000, p. 13

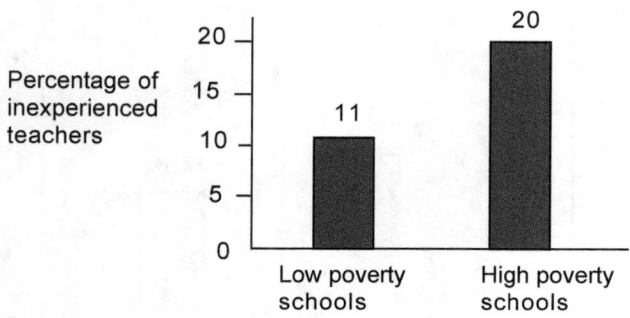

Figure 8. Poor students get more inexperienced teachers.
Source: Mayer, Mullens, & Moore, 2000, p. 13.

Access to qualified teachers in mathematics

The great majority of the least prepared teacher recruits are very frequently placed in under-resourced, hard-to-staff schools serving high proportions of low-income and minority students in central cities and poor rural areas. Students who most need skilled teachers are least likely to have well prepared teachers, thus magnifying inequalities (Darling-Hammond, 2001). For example, in California, the percentage of underprepared teachers in mathematics rises as the percentage of students of minority groups increases (see Figure 9) (Esch, Chang-Ross, Guha, Humphrey, Shields, Tiffany-Morales, Wechsler, & Woodworth, 2005).

More classes in high poverty schools and high minority schools are taught by out-of-field teachers, that is, by teachers who do not have at least a minor in the subject area they teach (Figure 10). Classes in high schools and middle schools with high percentages of Latino and African-American students are more likely to be taught by teachers who lack even a minor in the subject area than classes in schools with low percentages (Wilkins et al., 2006). This is a problem that has existed for quite some years. There was no progress in reducing the percentage of classes taught by of out of field teachers between 1994 and 2000 (Jerald, 2002). On the contrary, there was a slight increase, but this increase does not affect all students equally. While the percentages in low-poverty and low-minority schools remained essentially unchanged, the percentages of out of field classrooms in high-poverty and high-minority schools increased significantly (Figure 11).

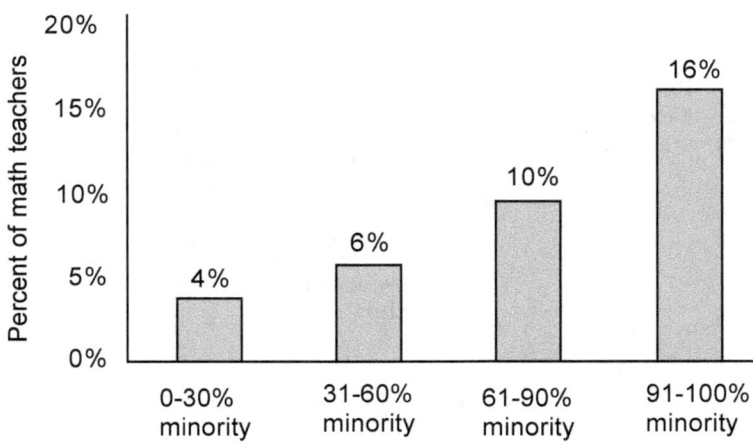

Figure 9. Distribution of underprepared teachers with a math assignment by school level percentage of minority students.
Source: Esch et al., 2005, p. 75

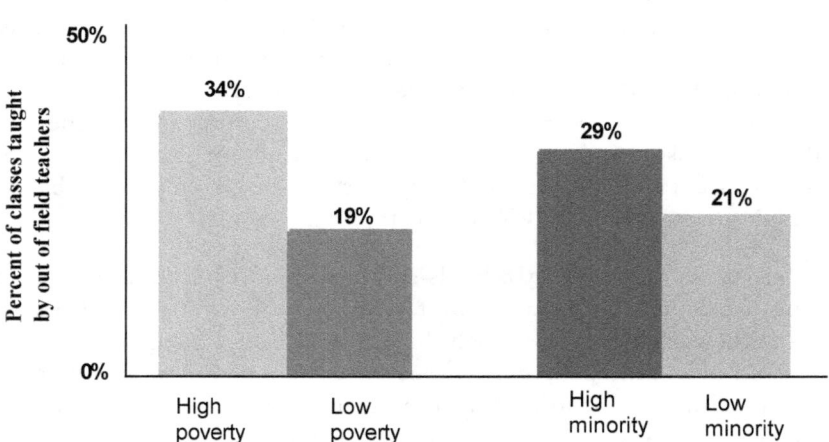

Figure 10. More classes in high poverty, high minority schools are taught by out-of-field teachers.
Source: Richard M. Ingersoll, University of Pennsylvania, as cited by Haycock, 2006, slide 61

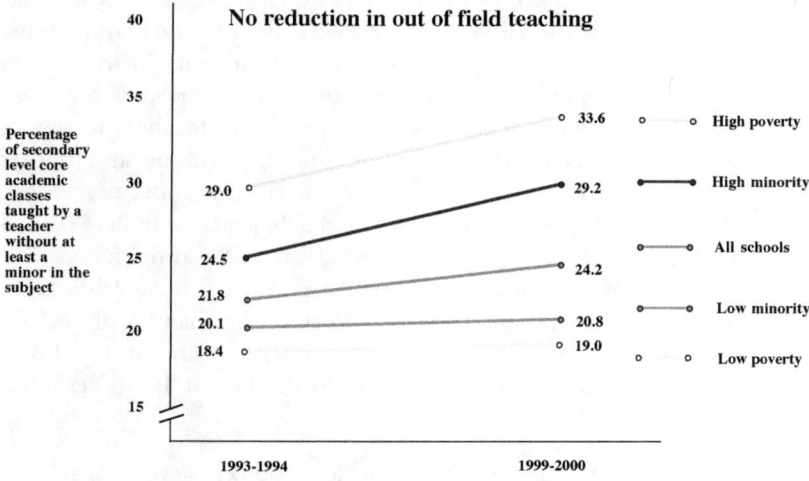

Figure 11. Changes in the percentages of classes taught out of field.
Source: Richard M. Ingersoll, University of Pennsylvania, as cited by Jerald, 2002, p. 5

To solve the problem of out of field teaching, action is needed at the school and district levels. Part of the problem is due in some places to a shortage of teachers in science and mathematics. However, a considerable part of the problem, about half (Jerald, 2002) could be solved with the present cadre of teachers by assigning teachers to teach in their field of expertise. At present, teachers in disadvantaged schools "are far more likely to be miss-assigned than are those in advantaged schools" (Ingersoll, 2002, p. 17).

Access to well-prepared teachers in general

Preparation in the content area and experience are not the only ways to measure teacher quality. The Illinois Education Research Council used five measures of teacher quality to define an overall "index" for schools, called the Teacher Quality Index (TQI). The five attributes they included are the percentage of teachers with degrees from more competitive colleges, the percentage of teachers with provisional or emergency credentials, the percentage of teachers who did not pass the Basic Skills tests their first time, the percentage of teachers with less than four years of experience, and the average ACT (American College Test) composite score of teachers in the school (Peske & Haycock, 2006). The researchers assigned to each school a Teacher Quality Index rating. Then they ranked all schools according to this rating and divided them into quartiles. Thus, schools in the top quartile had more teachers who were stronger on the measures included in the index than schools in the lower quartile. Then they analyzed patterns of distribution of qualified teachers according to the demographics of the student population. The data revealed that students in the schools with the highest concentrations of poverty and minority populations were assigned teachers who were qualitatively different from teachers in schools with low concentrations of students from minority groups and poverty (Figures 12 and 13). Large numbers of schools with the most percentages of students of minority groups had very low teacher quality indices. Of the schools with high percentages of minority students, 61% of them had TQIs in the lowest 10% of the state. Eighty-eight percent of those schools had TQIs in the bottom quartile of the state. On the other hand, only 11% of the schools that had the fewest minority students were in the lowest quartile, and only 1% of these schools were in the lowest 10% (Peske and Haycock, 2006).

The distribution of schools with high and low indices were similar when grouping schools by income. Eighty-four percent of the schools with the largest concentration of low-income students had indices in the bottom quartile in teacher quality, and 56% of those schools fell in the bottom 10% of the state. Only 1% percent of schools in the state (i.e., three schools total) serving high concentrations of poverty students had indices in the highest quartile of the state. In contrast, 46% of the schools with the lowest percentages of low-income students had a teacher quality index in the top quartile, and only 5% had an index in the lowest quartile (Figures 14 and 15) (Peske & Haycock, 2006, p. 7-8).

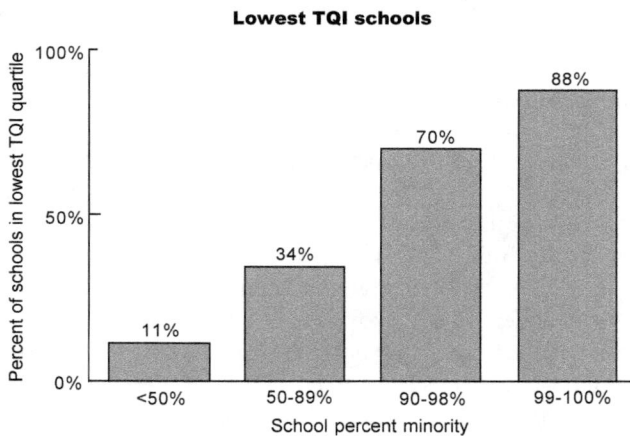

Figure 12. Low teacher quality index and percent minority Source: Peske & Haycock, 2006, p. 7

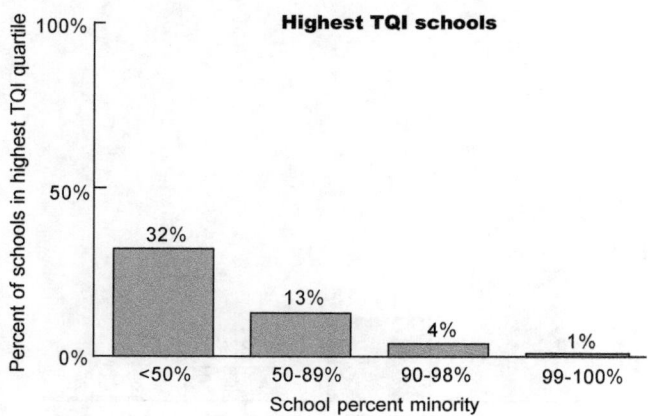

Figure 13. High Teacher quality index and percent minority
Source: Peske & Haycock, 2006, p. 7

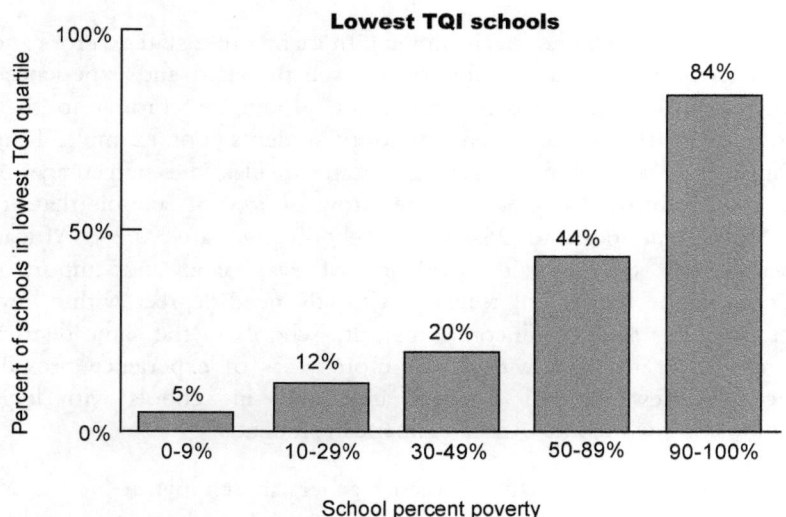

Figure 14. Low teacher quality index and percent poverty

Source: Peske & Haycock, 2006, p. 7

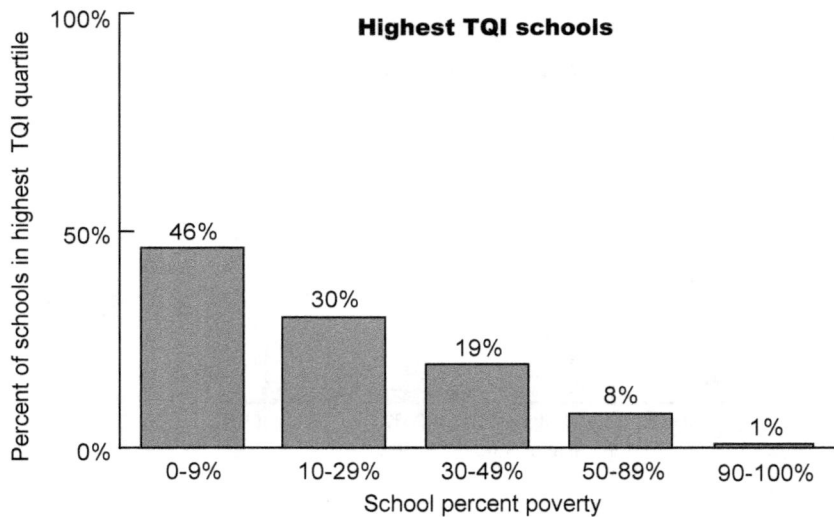

Figure 15. High teacher quality index and percent poverty
Source: Peske & Haycock, 2006, p. 7

The case of Illinois is by no means unique. In many other states, across and within school districts the distribution of well prepared and experienced teachers is not equitable when schools are grouped according to their concentrations of low-income and minority students. For example, Kain and Singleton (1996) found that in Greater Dallas the percentage of teachers with advanced degrees declines from 37.7% for schools that are 6.9% African American to 26.4% for schools that are 92.2% African American. Such data provide evidence of systematic and important differences in the fraction of teachers with advanced degrees within low-income, minority and high-income, majority schools in the same district. Their results for teachers with 20 or more years of experience provide evidence that fewer senior teachers are found in schools with large proportions of low-income, African American students.

There are also stark contrasts between teachers in schools in New York City and the rest of the state. Thirty-one percent of teachers in the public schools of the city failed the main certification exam at least once, whereas for the rest of the state the figure was only 5%. About 47% of teachers in the city who took the math state certification test failed at least once, compared with about 21% of teachers elsewhere. Twenty-seven percent of city teachers who took the state test for elementary teaching skills failed, compared with 3% elsewhere (Goodnough, 1999).

In California, the lowest performing schools have the highest percentages of underprepared and novice teachers. In 2005-06, in schools in the lowest achievement quartile of the state's Academic Performance Index, 21% of teachers were underprepared, novice, or both, whereas in the highest achievement schools it was 12% of teachers (Guha, Campbell, Humphrey, Shields, Tiffany-Morales, & Wechsler, 2006). The distribution of underprepared and novice teachers follows a similar pattern when using performance on the California High School Exit Exam. Thirty-one percent of teachers in schools with the lowest passing rates in the mathematics section were underprepared or novice in 2006, whereas in schools with the highest passing rates it was 17% (Guha et al., 2006). In 2005-2006 in high-minority middle and high schools, 16% of mathematics teachers were underprepared, compared with 4% in low minority schools. The distribution using the school's Academic Performance Index follows a similar pattern. In low performing middle and high schools, 18% of mathematics teachers were underprepared, compared with 5% in high performing schools (Guha et al., 2006).

There are also differences in distribution of qualified teachers within schools. Students in low tracks are more likely than students in high tracks to have a teacher without a minor or major in mathematics (Figure 16) (Ingersoll, 1999).

As Haycock (2000) points out, "No matter how you measure teacher qualifications—licensed vs. unlicensed, in- vs. out-of- field, performance on teacher licensure exams, or even actual effectiveness in producing learning gains—low-income and minority youngsters come up on the short end" (p. 1). To guarantee that the students with the most needs also get their fair share of strong teachers requires action at the school, district, state, and federal levels.

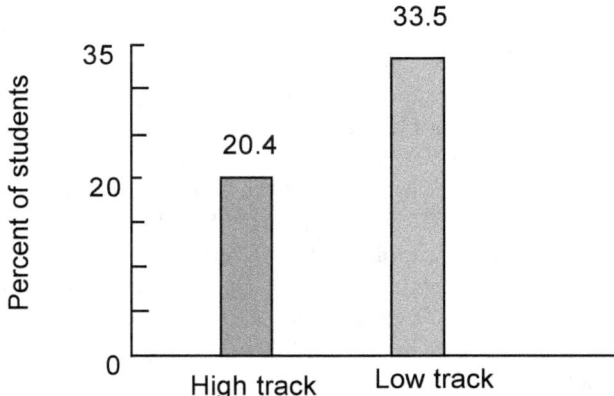

Figure 16. Students in lower track more likely to get less qualified teachers
Source: Ingersoll, 1999, p. 30

Gaps in opportunity due to low expectations

Low expectations can manifest themselves and be detrimental to students in several different ways. Low expectations often result in self-fulfilling prophecies. Once placed in the low tracks it is very difficult for students to move to higher tracks. Academic talent does not flourish and developed unless it is nurtured. Students who are not given the opportunity to learn challenging and interesting mathematics and instead are subject to rote learning of remedial mathematics are often turned off from mathematics and do not take mathematics beyond the minimum required, thus loosing options to mathematically oriented careers. Students who do not have access to advanced courses in mathematics in school are placed at a disadvantage when competing for a place at the best institutions of higher education.

Often students from a cultural background different from that of the teacher are put in situations where the teacher assumes deficits in the students, rather than locating and teaching to their strengths. Teachers may assume that the failure of a student to thrive intellectually is due to a deficit in the student rather than a deficit in teaching. As a consequence teachers may be teaching less when they should be teaching more (Delpit, 1992). Many mathematics teachers fail to use the culture of African American students in instruction. This results in the school culture being alienating to many African American students and being inconsistent with their cultural experiences, and also their dreams, hopes and struggles (Malloy, 1997, p. 27). The same could be said for many Latino students (Getz, 1997).

Lower expectations, ineffective teaching, and reinforced stereotypes often result from a sense of helplessness of the teacher based on teacher's beliefs of what students are capable of doing and the kind of support they receive at home (Irvine and York, 1993). Low expectations lead in turn to fewer opportunities for students to learn more challenging and advanced mathematics.

Different expectations for different students are often reflected in the ways teachers teach and test. Teachers who had at least 60% African American or Latino students in their classrooms were far more likely to spend classroom time using multiple-choice testing and other means of testing low-level cognitive objectives than teachers who had a majority of White students in their classrooms (Madaus, West, Harmon, Lomax, & Viator, 1992).

Giving high grades for work that in other schools would earn lower grades (see Figure 17) or would be more appropriate for younger students is another form in which low expectations are manifested. This can have a devastating effect later. A student may earn all As in her school and not realize how much farther behind she is with respect to her peers in other schools. It is not until such student has to compete with students from other schools to enter a school in high demand that she becomes fully aware how inadequate her preparation was.

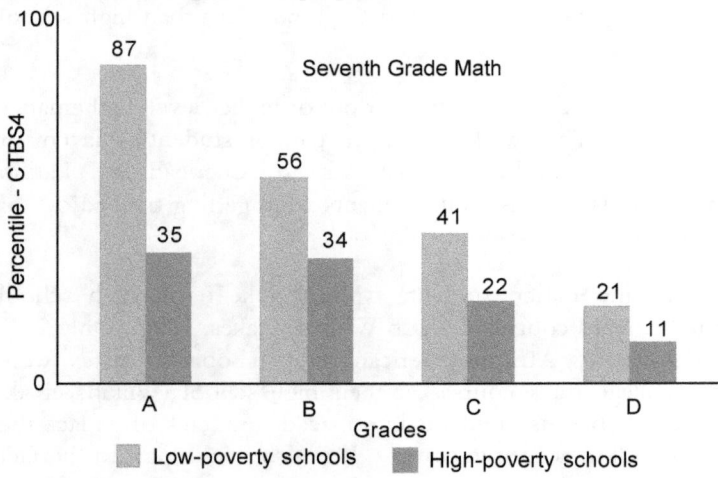

Figure 17. Students in poor schools receive As for work that would earn Cs in affluent schools.
Source: *Prospects* (Abt Associates, 1993), as cited by Haycock, 2006, slide 56.

African American and Latinos are more likely than Whites to be placed in low-ability and remedial classes or in special education programs and so they are more likely to learn fewer topics and skills (Oakes, 1990). African American and Latino students are also less likely to be identified as capable learners and placed in enriched or accelerated programs. For example, in 2000, 32% of White 8th-graders were in what teachers considered high ability classes, but only 16% of Latino and 16% of African American 8th-graders were in such classes (Strutchens, Lubienski, McGraw, & Westbrook, 2004). Consequently, African American and Latino students are likely to have fewer opportunities to learn science and mathematics than their White peers (Oakes, 1990, p. 160).

Enrollment of African American and Latino students is significantly lower than that of European American students in 8th-grade courses that to a great extent determine whether students will have the opportunity to take advanced mathematics (pre-calculus and calculus) before they graduate from high school. Only 49% of Latinos and 47% of African American students have taken pre-algebra or algebra in 8th grade compared to 68% of European American students (Strutchens et al., 2004).

Unfortunately it is not unusual to see African American and Latino students placed in low tracks even in cases when they scored as well as or better than their White or Asian American peers on standardized tests (Education Trust, 1996, as cited by Love, 2002, p. 258). Often African American and Latino are tracked out of advanced mathematics courses based on false assumptions. For example, Love (2002, p. 3) quotes an urban high school mathematics teacher:

We thought we were tracking students in or out of higher-level mathematics courses by their ability. Then we looked at the data on student achievement on standardized tests. We learned that African American and Latino students who scored as high as white students were getting tracked out of college-level courses.

African American and Latino students typically take fewer high school science and mathematics courses than do Whites (Oakes, 1990). Only 22% of Latino and 25% of African-American high school graduates were enrolled in the college track courses at their high school (Wilkins et al., 2006). In part, this is because students in nonacademic track often lack the prerequisites to enroll in academic courses. It is also due in part to the fact that they are very often required to take fewer science and mathematics courses than are college-bound students (Oakes, 1990). However, as Oakes (1990) points out, tracking is not the only factor in differences in course-taking. In many schools with large numbers of low income students, even those students in college preparatory programs typically take fewer academic classes.

Students need the opportunity to take more advanced level courses in mathematics. Not surprisingly, students who take more advanced mathematics courses do better on tests (see Figure 18). However, participation in more advanced mathematics courses is uneven among groups of different ethnic backgrounds (Figure 19). We need to encourage and expose students from all backgrounds to opportunities to take more advanced courses. Oakes (1990) reports that girls and students from minority groups usually receive less encouragement and have fewer science- and math-related opportunities both in school and out than do White males. However, when girls and students from minority groups do receive encouragement and are exposed to opportunities, they show interest and participate.

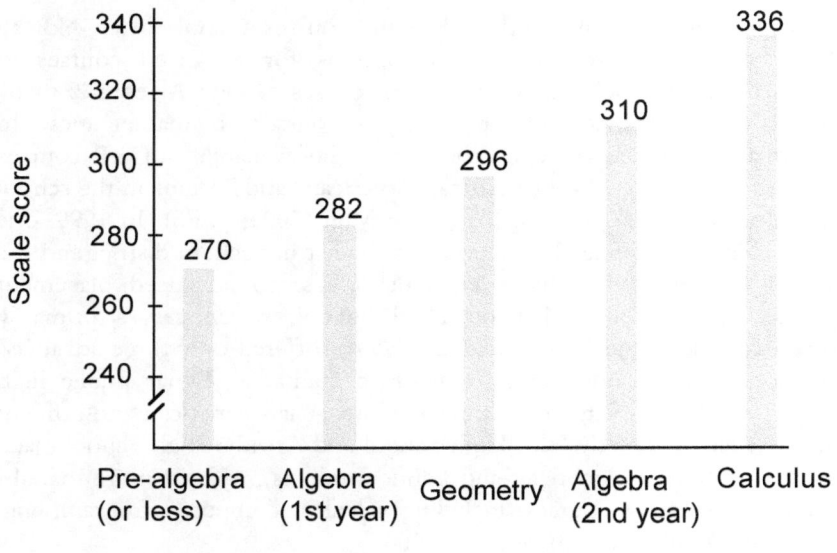

Figure 18. Mathematics scale scores, students age 17, by highest mathematics course taken: 2004
Source: Perie, Moran, & Lutkus, 2005, p. 57

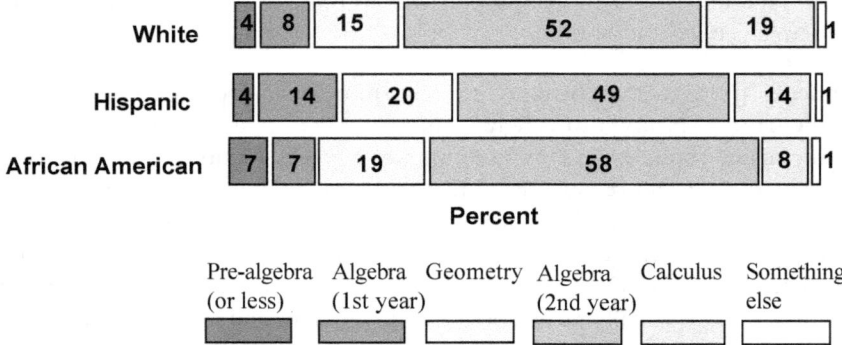

Figure 19. Percentage of students, by race/ethnicity and highest mathematics course taken: 2004
Source: Perie, Moran, & Lutkus, 2005, p. 58.

Low expectations are also reflected in the courses schools offer. Not all schools offer the same number of options for advanced courses in mathematics. Many schools do not offer courses beyond Algebra 2, many do not offer advanced placement (AP) courses in mathematics. In California, "regardless of high school size, the availability of AP courses decreases as the percentage of African Americans and Latinos in the school population increases" (Oakes, Joseph, & Muir, 2004, p. 75). In 1999, one student filed a statewide class-action suit against her school district and the state of California to achieve equitable access to advanced placement courses. Her school, Inglewood High School, which serves primarily African American and Latino students (99%), offered only three advanced placement courses, none in science or mathematics. Other public high schools, which serve large numbers of White, Asian American and affluent students such as Beverly Hills High School and Arcadia High School (each with 9% of African American and Latino students), offered more than 14 advanced placement courses, including calculus, computer programming, and physics (Oakes, Joseph, & Muir, 2004).

Gaps in opportunities to receive equal funding

Inadequate funding for schools often results in conditions that are detrimental for students and teachers; such as school buildings in disrepair, large numbers of students in each classroom, general overcrowding in the school, and insufficient or inadequate instructional materials (reference?). Limited funding also puts some districts at a disadvantage when competing for experienced and highly qualified teachers. "Some educators manage— by ingenuity, resourcefulness, and sheer force of will—to get very high achievement for students in high-poverty and high-minority schools despite egregious funding gaps. But it is undeniable that in the aggregate poor children have fewer opportunities in public schools in most states because they have fewer resources available to them" (Education Trust, 2005, p. 2).

Gaps across districts

In many places in the U.S., school funding is based mainly on local property taxes and revenues (Howell and Miller, 1997). School districts with a large number of well-to-do people have more funds per student than school districts with a large number of people in poverty. Thus students of poverty tend to go to schools with less funds. In many places also, a large proportion of African American and Latino students live in districts with less funding available. According to NAEP data from 2000, only 3% of White 8th-graders are in schools where more than 75% of students qualify for free or reduced-price lunch, whereas 34% of African American and 34% of Latino 8th-graders are in such schools. The majority of White 8th-graders attend schools with less poverty. While 64% of White 8th graders attend schools with less than one quarter of the students being eligible for free or reduced price lunch, only 15% of African American and 25% of Latino 8th graders do so (Strutchens et al., 2004 p. 281).

There can be no real equal opportunity to learn as long as the "savage inequalities" in our schools continue to exist (Kozol, 1991). In many places, schools are highly segregated. For example, in the Bay Area in California, although 28% of the students are White, 65% of White students go to White-majority schools (Oakes, Rogers, Silver, Horng, & Goode, 2004). Schools are not only segregated, they are also unequal. School districts that educate the greatest number of African-American and Latino students receive less local and state money to educate them than the districts serving the fewest number of minority students (Wilkins et al., 2006). Teachers in schools with the highest concentrations of low-income students are less likely to obtain the resources they need. Barton, Coley, and Goertz (1991) report that while more than 80% of the teachers in schools with middle- to upper-SES students received all or most of the materials they requested for instructional purposes, only 41% of teachers in schools with mainly low-SES students received all or most of the instructional materials they requested.

In California, schools where 90% or more of the students are African American, American Indian, Latino, Filipino, and/or Asian are much more likely than majority-White schools to have serious problems such as instructional materials that are inadequate, facilities in disrepair, and overcrowding. Forty-two percent of schools with high concentrations of African American, American Indian, Latino, Filipino, and/or Asian students have such serious problems compared to only 7% of White-majority schools. In Los Angeles County, 63% of schools with large proportions of African American, American Indian, Latino, Filipino, and/or Asian students suffer such serious problems (Oakes, Rogers, Silver, Horng, & Goode, 2004).

In some cases, legal actions are necessary to redress disparities in funding. For example, the difference of per-student expenditure in New York City and other parts of the state was so big that the state was sued to allocate funds for students in more equitable ways. Recently, the state's highest court ruled that the city should be allocated at least $1.93 billion more per year. Although this is far less than the $4.7 billion set by a lower court (Herzenhorn, 2006), it is a clear indication that the funds allocated for students in the city were not sufficient.

The funding inequities of New York City are not unique by any means. In many urban areas there are huge disparities in per student spending between districts serving large numbers of Latino and African American and low-income students and districts in the suburbs with low concentrations of African American and Latino students and low numbers of students who qualify for free or reduced price lunch. Table 1 shows per student spending for two districts in each of several metropolitan areas (Kozol, 2005). It also gives the percentage of students who are African American or Hispanic, and the percentage of low income students for each district. The pattern is unmistakable. In each metropolitan area, the higher the percentage of Latino and African American students, the lower the per student spending. The differences are also impressive. In some cases, the per-student spending in a low-minority district is twice as much as in the neighboring district with large numbers of African American or Latino students.

Table 1. Per student spending in several metropolitan areas 2002-2003

Adapted from Kozol, 2005

Metropolitan area	School District	Spending per student	% Hispanic + African American	% Low income
Chicago area	Highland Park and Deerfield (HS)	$17,291	10	8

	Chicago	$8,482	87	85

Philadelphia area	Lower Merion	$17,261	9	4
	Philadelphia	$9,299	79	71

Detroit area	Bloomfield Hills	$12,825	8	2
	Detroit	$9,576	95	59

Milwaukee area	Maple Dale - Indian Hill (K-8)	$13,955	20	7
	Milwaukee	$10,874	77	76

Boston area	Lincoln (K-8)	$12,775	19	11
	Lawrence	$7,904	86	69

New York City area	Manhasset	$22,311	9	5
	New York City	$11,627	72	83

Districts with less resources often are not able to compete in teachers' salaries with wealthier districts. At the time the state of New York was sued for inequitable funding, the plaintiffs pointed out that the starting salary for New York City teachers was about 25% less than starting salaries in wealthy suburban counties (Goodnough, 1999). Who would blame a teacher who needs to tend for her or his family for moving to another school district where the pay is better?

Gaps in opportunity to receive equal funding within districts

Inequities in funding across districts are easy to see just by visiting school buildings in wealthier districts and in poorer districts. However, less known is the fact that the problem of unequal funding exists also *within* districts. Schools with a larger proportion of minority or low income students within the same district often have a larger proportion of inexperienced teachers. Wiener (2006) describes two schools in the same district in San Diego. One school has 55% White students, 32% students on free or reduced price lunch, and an Academic Performance Index of 808 (the statewide performance target is 800). The other school has 79% Latino and African American students, 75% students receive free or reduced price lunch, and the Academic Performance Index is 648. The average pay for teachers at the first school was $6,800 higher than at the second. This difference in salaries happens because as more experienced teachers migrate from one school to another they take their higher salaries with them. Teachers with more experience tend to migrate to schools with a larger proportion of European American students, to schools with less poverty, and schools that perform better overall in state mandated tests (Wiener, 2006).

The San Diego example is not an isolated incident. In many urban districts there are huge differences in average salaries for teachers from one school to another. This inequity is not transparent due to the fact that urban districts calculate school budgets using average teacher costs. Thus a school with a staff consisting of mainly senior staff with higher salaries does not appear in the official budget as receiving more money than another school that is staffed mainly by beginning teachers with lower salaries. The hidden differences can be staggering. For example, in Baltimore City Schools the real costs were very different from the official budgets. While the district average was $47,178, in one school the average salary of the teachers was only $37,618, and at another school the average was more than $57,000 (Roza & Hill, 2004). Thus, the teacher expenditure per student is very different from school to school. In Austin, the distribution of non-categorical spending per student is not equitable across schools. Teachers who teach at schools with the greatest enrollment of low-income students receive just 85% of the district salary average, while those teachers at schools with the smallest need receive 108% of the district average (Roza, Miller, & Hill, 2005).

Unfortunately, the schools that are thus shortchanged are schools with large numbers of low income children. The schools that benefit from this budgeting system are schools with larger proportions of upper income students. By using average costs for the school budgets, districts mask the fact that they are taking away from the poor to benefit the rich. Federal programs that allocate funds that are meant to supplement and not supplant allow this practice. The money received from the federal government is meant to be used to give additional resources to students of poverty, not to replace money that had been allocated to serve other purposes. For many years, Title I legislation allowed districts to use average salary figures when comparing expenditures among schools (Roza & Hill, 2004, p. 212). "Districts were henceforth allowed to maintain major inequities in school funding, as long as these were driven by teacher allocation" (Roza & Hill, 2004, p. 216). The biggest part of school budgets comes from teachers salaries, typically more than 80% of the school allotment (Roza & Hill, 2004), so inequities in teacher salaries across schools amount to large inequities in per student expenditure.

We clearly have huge opportunity gaps. By focusing on the gaps of opportunities, it becomes clear that the achievement gap is just one of the symptoms. To solve the problem, we need to address the causes not just the symptoms. As Lee Bollinger, President of Columbia University states:

> The achievement gap that exists in American education is not a gap in ability, but a gap in resources and a gap in expectations. We know that students from all backgrounds can succeed at the highest levels of education, when they are given the support they need to succeed—the support that is regularly given to the students from the top income brackets. (Quoted by Scatton, Coley, & McBride, 2006, p. 2)

Final comments

As we have seen, some important groups of students have considerably less opportunities to learn in our schools than other groups. We need to be aware of the structural inequities that exist across school districts, as well as within districts, schools, and classrooms. Often, inequities are assumed to reflect a hierarchy of competence. Students who had more opportunities are assumed to be more capable or have more aptitude to learn than other students who did not have as many opportunities.

Talking about achievement gaps without mentioning the opportunity gaps that cause them is an invitation to look at the students who lag behind and perform poorly through deficit models to try to "explain" low performance in terms of factors such as cultural differences, poverty, low educational level of the parents, or experiences these students lack. As Khisty (1995) points out, we can look at the problem in a fundamentally different way and identify the instructional and organizational decisions that harm some of our students. By describing the problem in terms of opportunity gaps, we focus our attention on some of the important things that some of our students are not receiving at school. This focus makes it more clear what the actions are we need to take to guarantee that all students do indeed have the same opportunities to receive a high quality education in school.

As mentioned before, the opportunity gaps described in this chapter—less access to experienced and well qualified teachers, facing low expectations, and inequities in per student funding—are not the only challenges that many Latino students face in our schools. Other issues such as language (Cuevas, 1990), cultural differences, and bias in testing are also important. Latinos constitute a widely diverse cluster of groups and some among them, for example, recent immigrants, have special needs. They do not only face the challenge of learning mathematics in a new language, but also important differences in the way they or their parents learned mathematics in their countries of origin (Perkins & Flores, 2002). It is important that schools do more to accommodate the special needs of these groups of students and this will require additional expertise, resources, and funding. However, a great deal would be gained if the opportunities to learn for Latinos and other groups that have been systematically underserved by the educational system were the same as for other students. A great deal would probably be advanced towards having the same achievement across groups by providing real equity in opportunities to learn within schools and across schools.

As many examples across the nation show, given the opportunity, students from any cultural or ethnic background and any socioeconomic level can excel. Kitchen, DePree, Celedón-Pattichis, & Binkerhoff (2007) found three salient characteristics in all the schools they studied that were highly successful educating students of poverty, especially in mathematics. These characteristics are "(a) high expectations and sustained support for academic excellence, (b) challenging mathematical content and high-level mathematics instruction that focused on problem solving and sense making (as opposed to rote instruction), and (c) the importance of building relationships" (p. xiv). The opportunity to have access to schools and teachers with such characteristics should be the rule rather than the exception for students from all groups. Concerted actions are needed at the classroom, school, district, state, and federal levels to guarantee that all students receive the same opportunities to learn mathematics.

Of course, achievement and participation are not the only aspects that are important for equity in mathematics education. The practice of mathematics "to analyze, reason about, and especially critique knowledge and events in the world" (Gutiérrez, 2002, p. 158)) is also crucial to achieve an equity agenda. Developing rich mathematical activities based on the wealth of knowledge of the families and communities of the students is another avenue for equity (Civil, 2004). Much more than access to a traditional curriculum is needed to prepare students to develop a sense of agency and "write the world" with mathematics (Gutstein, 2006).

Equity of opportunities in schools is but one of the aspects needed to achieve equity of opportunities in society. As Ladson-Billings (2006) points out, injustices of the past still affect us today, a situation she has described as the "education debt." Today's children whose parents were shortchanged in school in the past will probably face greater challenges in school and in life than children whose parents had all the opportunities and advantages. The least we can do is to make sure that this education debt does not keep increasing for present and future generations.

References

Barton, P. E., Coley, R. J., & Goertz, M. E. (1991). *The state of inequality.* Princeton, NJ: Policy Information Center. Educational Testing Service.

Civil, M. (2004). Building on community knowledge: An avenue to equity in mathematics education. Retrieved March 18, 2007 from http://math.arizona.edu/~cemela/spanish/content/workingpapers/BuildingCommunityKnowledge.pdf

Cuevas, G. (1990). Increasing the achievement and participation of language minority students in mathematics education. In T. J. Cooney & C. R. Hirsch (Eds.), *Teaching and learning mathematics in the 1990s* (pp. 159-165). Reston, VA: National Council of Teachers of Mathematics.

Darling-Hammond, L. (2000). Teacher quality and student achievement: A review of state policy evidence. *Educational Policy Analysis Archives, 8*(1).

Darling-Hammond, L. (2001). The challenge of staffing our schools. *Educational Leadership, 58*(8), 12-17.

Delpit, L. D. (1992). Education in a multicultural society: Our future's greatest challenge. *Journal of Negro Education, 61*(3), 237-249.

Diversity in Mathematics Education Center for Learning and Teaching. (2007). Culture, race, power, and mathematics education. In F. K. Lester (Ed.), *Second handbook of research on mathematics teaching and learning* (Vol. 1, pp. 405-433). Charlotte, NC: Information Age Publishing.

Education Trust. (1996). *Education watch: The Education Trust community data guide.* Washington, DC: Education Trust.

Education Trust. (2005). *The funding gap 2005: Low-income and minority students shortchanged by most states.* Washington, DC: Education Trust. Retrieved January 3, 2007 from http://www2.edtrust.org/NR/rdonlyres/31D276EF-72E1-458A-8C71-E3D262A4C91E/0/FundingGap2005.pdf

Esch, C. E., Chang-Ross, C. M., Guha, R., Humphrey, D. C., Shields, P. M., Tiffany-Morales, J. D., Wechsler, M. E., & Woodworth, K. R. (2005). *The status of the teaching profession 2005.* Santa Cruz, CA: The Center for the Future of Teaching and Learning.

Getz, L. M. (1997). *Schools of their own: the education of Hispanos in New Mexico.* Albuquerque: University of New Mexico Press.

Goodnough, A. (1999, November 16). City's teachers perform poorly on state exams. *New York Times,* p. 1.

Greene, J. P., & Foster, G. (2004). The teachability index: Can disadvantaged students learn? Education Working Paper No. 6 (September). Retrieved December 3, 2006 http://www.manhattan-institute.org/html/ewp_06.htm

Guha, R., Campbell, A., Humphrey, D., Shields, P., Tiffany-Morales, J., & Wechsler, M. (2006). *California's teaching force 2006: Key issues and trends.* Santa Cruz, CA: The Center for the Future of Teaching and Learning.

Gutiérrez, R. (2002). Enabling the practice of mathematics teachers in context: Toward a new equity research agenda. *Mathematical Thinking and Learning, 4*(2&3), 145-187.

Gutstein, E. (2006). *Reading and writing the world with mathematics.* New York: Routledge.

Haycock, K. (2000). Honor in the boxcar: Equalizing teacher quality. *Thinking K-16, 4*(1), 1-2.

Haycock, K. (2006). Improving achievement and closing gaps, Pre-K through college. Retrieved December 10, 2006 from http://www2.edtrust.org/EdTrust/Product+Catalog/recent+presentations.htm

Heffter, E. (2006, November 30). WASL achievement gap costly to fix, officials say. *The Seattle Times.* Retrieved December 10, 2006, from http://seattletimes.nwsource.com/html/education/2003454369_gap30 m.html

Herzenhorn, D. M. (2006, November 21). New York cuts aid sought for city schools. *New York Times,* p. 1.

Hiebert, J., & Grouws, D. A. (2007). The effects of classroom mathematics teaching on students' learning. In F. K. Lester (Ed.), *Second Handbook of Research on Mathematics Teaching and Learning* (Vol. 1, pp. 371-404). Charlotte, NC: Information Age Publishing.

Howell, P. L., & Miller, B. B. (1997). Sources for funding for schools. *The Future of Children, 7*(3), 39-50.

Ingersoll, R. M. (1999). The problem of underqualified teachers in American secondary schools. *Educational Researcher, 28*(2), 26-37.

Ingersoll, R. M. (2002). *Out-of-field teaching, educational inequality, and the organization of schools: An exploratory analysis.* Seattle, WA: Center of the Study of Teaching and Policy, University of Washington.

Irvine, J. J., & York, D. E. (1993). Teacher perspectives: Why do African-American, Hispanic and Vietnamese students fail? In S. W. Rothstein (Ed.), *Handbook of schooling in urban America* (pp. 161-173). Westport, CT: Greenwood Press.

Jerald, C. D. (2002). *All talk, no action: Putting an end to out-of-field teaching.* Retrieved December 5, 2006 from http://www2.edtrust.org/NR/rdonlyres/8DE64524-592E-4C83-A13A-6B1DF1CF8D3E/0/AllTalk.pdf

Kain, J. F., & Singleton, K. (1996). Equality of educational opportunity revisited. *New England Economic Review* (May/June), 87-111.

Khisty, L. L. (1995). Making inequality: Issues of language and meaning in mathematics teaching with Hispanic students. In W. G. Secada, E. Fennema & L. B. Adajian (Eds.), *New directions for equity in mathematics education.* New York: Cambridge University Press.

Kitchen, R. S., DePree, J., Celedón-Pattichis, S., & Brinkerhoff, J. (2007). *Mathematics education at highly effective schools that serve the poor: Strategies for change.* Mahwah, NJ: Lawrence Erlbaum Associates.

Kossan, P. (2006, December 30). Schools appeal poor rankings. *The Arizona Republic,* p. A1.

Kozol, J. (1991). *Savage inequalities: Children in America's schools*. New York: Crown.

Kozol, J. (2005). *The shame of the nation: The restoration of apartheid schooling in America*. New York: Three Rivers Press.

Ladson-Billings, G. (2006). From the achievement gap to the education debt: Understanding achievement in U.S. schools. *Educational Researcher, 35*(7), 3-12.

Love, N. (2002). *Using data / getting results: A practical guide for school improvement in mathematics and science*. Norwood, MA: Christopher-Gordon Publishers.

Madaus, G. F., West, M. M., Harmon, M. C., Lomax, R. G., & Viator, K. A. (1992). *The influence of testing on teaching math and science in grades 4-12*. Chestnut Hill, MA: Boston College, Center for the Study of Testing, Evaluation, and Educational Policy. (ERIC No. ED370772.)

Malloy, C. E. (1997). Including African American students in the mathematics community. In J. Tentracosta (Ed.), *Multicultural and gender equity in the mathematics classroom: The gift of diversity* (pp. 23-33). Reston, VA: National Council of Teachers of Mathematics.

Mayer, D. P., Mullens, J. E., & Moore, M. T. (2000). *Monitoring school quality: An indicators report, NCES 2001-030*. Washington, DC: National Center for Education Statistics. Retrieved December 17, 2006 from http://nces.ed.gov/pubs2001/2001030.pdf

National Council of Teachers of Mathematics. (2005). Closing the achievement gap. Reston, VA: National Council of Teachers of Mathematics. Retrieved November 19, 2006 from http://www.nctm.org/about/content.aspx?id=6350

National Governors' Association. (2005). *Closing the achievement gap*. Retrieved January 4, 2007, from http://www.subnet.nga.org/educlear/achievement/

Oakes, J. (1990). Opportunities, achievement and choice: Women and minority students in science and mathematics. *Review of Research in Education, 16*, 153-222.

Oakes, J., Joseph, R., & Muir, K. (2004). Access and achievement in mathematics and science: Inequalities that endure and change. In J. A. Banks & C. A. M. Banks (Eds.), *Handbook of Research on Multicultural*

Education (second ed., pp. 69-90). San Francisco: Jossey-Bass.

Oakes, J., Rogers, J., Silver, D., Horng, E., & Goode, J. (2004). *Separate and unequal 50 years after Brown: California's racial "Opportunity Gap"*. Los Angeles, CA: Institute for Democracy, Education, and Access, UCLA. Retrieved January 4, 2007 from http://idea.gseis.ucla.edu/publications/idea/images/brownsu2.pdf

Perie, M., Moran, R., & Lutkus, A.D. (2005). *NAEP 2004 Trends in Academic Progress: Three Decades of Student Performance in Reading and Mathematics*. Washington, DC: National Center for Education Statistics. Retrieved December 3, 2006 from http://nces.ed.gov/pubsearch/pubsinfo.asp?pubid=2005464

Perkins, I., & Flores, A. (2002). Mathematical notations and procedures of recent immigrant students. *Mathematics Teaching in the Middle School, 7*, 346-351.

Peske, H. G., & Haycock, K. (2006). *Teaching inequality: How poor and minority students are shortchanged on teacher quality*. Washington, DC: Education Trust. Retrieved December 18, 2006 from http://www2.edtrust.org/NR/rdonlyres/010DBD9F-CED8-4D2B-9E0D-91B446746ED3/0/TQReportJune2006.pdf

Roza, M., & Hill, P. T. (2004). *How within-district spending inequalities help some schools to fail*. Retrieved December 17, 2006, from http://muse.jhu.edu/journals/brookings_papers_on_education_policy/toc/pep2004.1.html

Roza, M., Miller, L., & Hill, P. (2005). *Strengthening Title I to help high-poverty schools: How Title I funds fit into district allocation patterns*. Retrieved December 18, from http://www.crpe.org/workingpapers/pdf/TitleI_reportWeb.pdf

Sanders, W. L., & Rivers, J. C. (1996). *Cumulative and residual effects of teachers on future student academic achievement*. Knoxville, TN: University of Tennessee Value-Added Research and Assessment Center.

Scatton, L. H., Coley, R. J., & McBride, A. (2006). Addressing Achievement Gaps: Leading the challenge of developing high-potential youth. *Policy Notes: News from the ETS Policy Information Center, 14*(1), 1-11.

Strutchens, M. E. (2000). Confronting beliefs and stereotypes that impede the mathematical empowerment of African American students. In M. E. Strutchens, M. Johnson, & W. Tate (Eds.). *Changing the faces of mathematics: Perspectives on African Americans* (pp. 7- 14). Reston, VA: National Council of Teachers of Mathematics.

Strutchens, M. E., Lubienski, S., McGraw, R., & Westbrook, S. K. (2004). NAEP findings regarding race/ethnicity: Students' performance, school experiences, attitudes and beliefs, and family influences. In P. Kloosterman, & F. K. Lester (Eds.) *Results and interpretations of the 1990 through 2000 mathematics assessments of the National Assessment of Educational Progress* (pp. 269-304). Reston, VA: National Council of Teachers of Mathematics.

Strutchens, M. E., & Silver, E. A. (2000). NAEP findings regarding race/ethnicity: Students performance, school experiences, and attitudes and beliefs. In E. A. Silver & P. A. Kenney (Eds.), *Results from the seventh mathematics assessment of the National Assessment of Educational Progress* (pp. 45-72). Reston, VA: National Council of Teachers of Mathematics.

Tate, W. F. (1997). Race-ethnicity, SES, gender, and language proficiency trends in mathematics achievement: An update. *Journal for Research in Mathematics Education, 28*(6), 652-679.

Wiener, R. (2006). Understanding opportunity gaps that are behind achievement gaps. UNC Symposium: High poverty schooling in America. Retrieved December 4, 2006 from http://www.law.unc.edu/pdfs/Wiener.pdf

Wilkins, A. & Education Trust staff. (2006) Yes we can: Telling truths and dispelling myths about race and education in America. Washington DC: The Education Trust. Retrieved November 20, 2006 from http://www2.edtrust.org/NR/rdonlyres/DD58DD01-23A4-4B89-9FD8-C11BB072331E/0/YesWeCan.pdf

22 THE MEAN AS THE BALANCE POINT: THOUGHT EXPERIMENTS WITH MEASURING STICKS[22]

Representativeness is one of the important aspects of the arithmetic average or mean of a set of data. Children's concepts of representativeness of averages include five aspects: Average as a mode, average as algorithm, average as reasonable, average as midpoint, and average as point of balance (Mokros & Russell, 1995). Of these, the mean as a balance point is the more sophisticated. According to Mokros and Russell, students who have the average as a point of balance approach are able to

- view an average as a tool for making sense of data;
- look for a point of balance to represent the data;
- take into account the values of all the data points;
- use the mean with a beginning understanding of the quantitative relationship among data, total, and average; they are able to work from a given average to data, from a given average to total, from a given total to data;
- break problems into smaller parts and find "submeans" as a way to solve more difficult averaging problems. (p. 26)

In this article we present some activities that will help students establish a connection between the mean and the balance point of a lever. Hardiman,

---

[22] Flores, A. (2008). The mean as the balance point: Thought experiments with measuring sticks. *International Journal of Mathematical Education in Science and Technology*, *39*(6), 741-748. Reprinted by permission of Taylor & Francis (http://www.tandfonline.com).

Well, and Pollatsek (1984) found that students can perform significantly better on problems related to weighted averages after having worked on ideas related to balance point. However, they state that an awareness of the analogy is an important condition for transfer, and also equally important is to understand the first principle sufficiently so that some "deep structure" can be transferred (p. 793). According to them, "if students have available a balance model of the mean, the benefit that can be derived in solving problems having to do with the mean will depend on the extent they understand the rules of balancing" (p. 795). If students do not have a good understanding of the balancing issues with a lever it is unlikely that they will be able to use those of the lever to build a better understanding of weighted averages and of the mean as a balance point. Therefore, we present first a discussion of the principles of the lever.

Preliminary discussion about the lever

If two children of equal weight want to balance on a see-saw, they have to sit symmetrically with respect to the fulcrum where the see-saw rotates. That is, they sit at the same distance from the point of balance, but in opposite directions (Figure 1). The point of balance is the midpoint. Two children of unequal weight want to balance on a see-saw (see figure 2). Where does the heavier child have to sit?

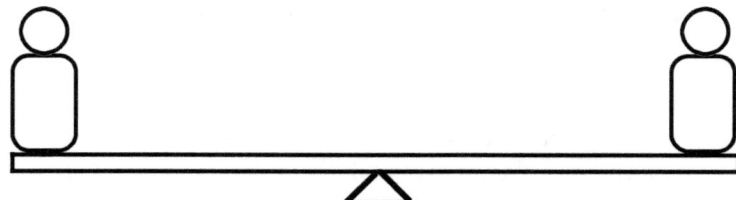

Figure 1. The balance point

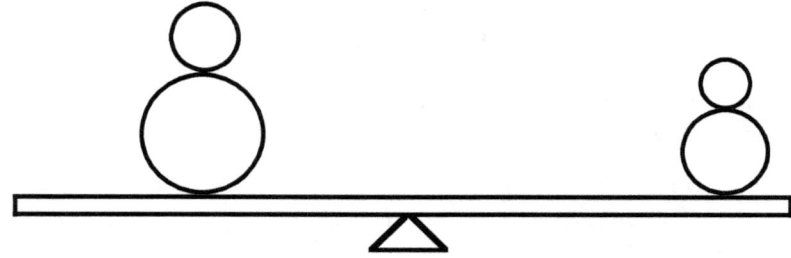

Figure 2. Two children of unequal weight want to balance.

The weight of one child on one side will tend to rotate the lever in one direction, while the weight of the other child will tend to rotate it in the opposite direction. This tendency to rotate will depend not only on the weight, but also on the distance of the weight from the fulcrum. The distance between the fulcrum and the weight is the *lever arm* (see figure 3).

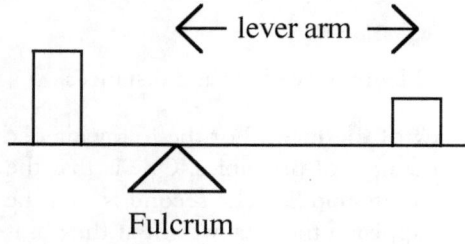

Figure 3. Lever arm

The *moment of force* or *torque* is defined as the product of the force (in this case the weight) by the distance to the fulcrum. The system will be in equilibrium when the clockwise torques cancel out the counterclockwise torques.

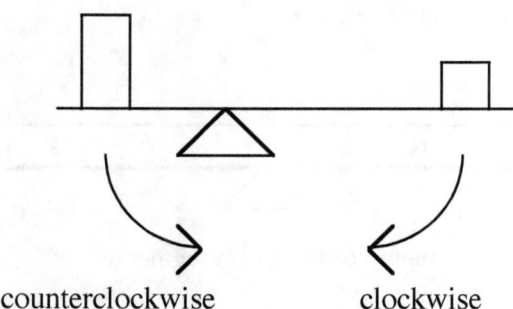

Figure 4. Canceling torques

The law of the lever. Let w_1 and w_2 be weights on a lever, and d_1 and d_2 be their respective distances to the fulcrum. If the lever is in equilibrium, then $w_1 d_1 = w_2 d_2$, that is, the clockwise moment is equal to the counterclockwise moment (see figure 5).

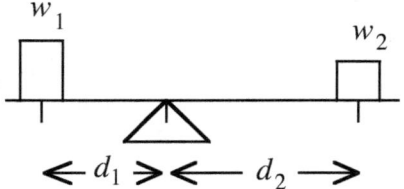

Figure 5. Weights and distances.

Derivation of the law of the lever. For the following derivation of the law of the lever we use a couple of principles. One is that the center of gravity of a uniform beam is in its middle. The second is that the action of a mass can be substituted by applying the total weight at the point where the center of gravity is. Another principle is that the weights of beam of uniform density with uniform thickness will be proportional to their lengths, so the weights of a beam of length a and a beam of length b will be ka and kb. Consider a beam of length $a + b$ that is balanced on a lever. If the beam is cut into two parts of lengths a and b, the system will still be in equilibrium. If we consider the left end of the lever to be at 0, the fulcrum will be at $\dfrac{a+b}{2}$, the center of gravity of the first beam will be at $a/2$, and the center of gravity of beam b will be at $a + b/2$.

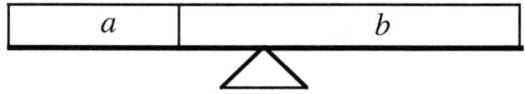

Figure 6. The beam segmented

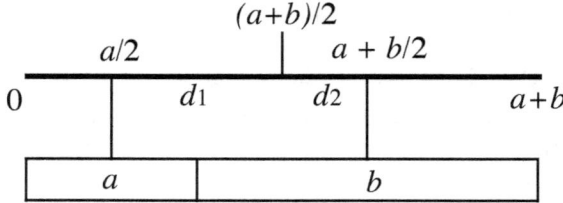

Figure 7. Actions at the centers of gravity

The distances of the centers of gravity to the fulcrum are given by

$$d_1 = \frac{a+b}{2} - \frac{a}{2} = \frac{b}{2}$$

$$d_2 = a + \frac{b}{2} - \frac{a+b}{2} = \frac{a}{2}$$

Thus $w_1 d_1 = ka \times \frac{b}{2}$ and $w_2 d_2 = kb \times \frac{a}{2}$, and we see that the moments of force are equal. Polya (1977) uses other approaches using basic principles to derive the general law of the lever.

Activities with a meter stick relating the mean to the balance point.

Activities using a meter stick and washers that relate the average with the balance point can be conducted with students (Flores and White 1989). However, in order for the activity to work well, the meter sticks and washers have to be carefully tested for balance, and materials need to be carefully prepared beforehand. Even a slight difference in weights or having the balance point off the center can cause confusion for students. Another approach is to consider an idealized situation where the lever is weightless (Flores 1988, 1995). For some students this approach may be too abstract. In this article we present yet another alternative, where measuring sticks and weights are still used to provide a concrete context, and some kinesthetic feeling, but the activities are conducted mainly as thought experiments.

Let students find the center of gravity of a meter stick. They can place the meter stick on top of the two separated hands and slide the hands towards the center. They will find that the balance point is at or very close to the 50 cm mark. If the meter stick were exactly 1 meter long, and the wood was of uniform density, the center of gravity, or balance point would be exactly at the midpoint. In what follows we will assume that the meter stick is balanced at its midpoint, that is, at the 50 cm mark.

The balanced meter stick. Students should imagine the meter stick balanced on a fulcrum placed at the 50 cm mark (Figure 9). The system will still be in equilibrium if we place equal weights on both sides of the lever at the same distance from the fulcrum (Figure 10). The total torque that would make rotate the lever clockwise should be the same as the total torque that would make rotate the lever counter clockwise. If there is a unit of weight at 20 and another unit of weight at 80, that the system is in equilibrium means that $(80 - 50) \times 1 = (50 - 20) \times 1$. From now on, we will consider directed distances in the computation of the torque to take into account the direction of rotation. With unit weights at 80 and 20, $(80 - 50) \times 1$ is the torque that will make rotate the lever clockwise, and the torque that tends to rotate the lever counter clockwise is $(20 - 50) \times 1$. For the system to be in equilibrium the sum of the torques should be zero, $(20 - 50) \times 1 + (80 - 50) \times 1 = 0$. Notice that 50 is the mean average of 20 and 80, $50 = \frac{20+80}{2}$.

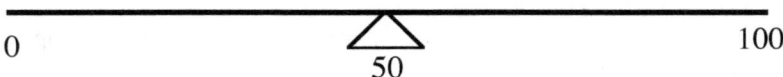

Figure 9. A balanced meter stick

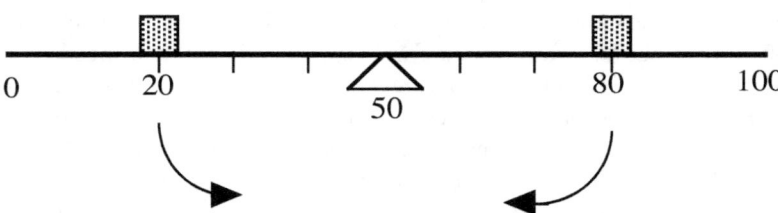

Figure 10. A balanced system

If we add a weight on top of the balance point, the balance will not be altered. In the same way, the average does not change if a value equal to the average is added to the data, $50 = \dfrac{20+80+50}{3}$.

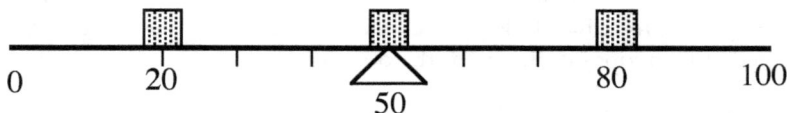

Figure 11. The equilibrium is maintained.

Another basic principle of the lever is that we can substitute two separate unit weights at different places by two joint weights at their midpoint without affecting the balance. The weights at 50 and 80 are substituted by two weights at 65 and the balance will be preserved. Similarly, the average of a set of data will still be the same if we substitute two data by two copies of their arithmetic average, $50 = \frac{20+65+65}{3}$.

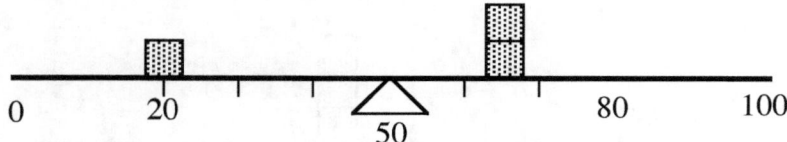

Figure 12. Still in equilibrium.

We can now do a reverse process and replace the two weights at 65 by two other weights placed so that 65 is their midpoint, for example at 60 and 70, and the balance will be maintained. Now we have two torques that tend to rotate the lever clockwise $(60 - 50)$ and $(70 - 50)$, and one torque that tends to rotate the lever counter clockwise $(20 - 50)$. The system is in equilibrium because $(20 - 50) + (60 - 50) + (70 - 50) = 0$. Notice that 50 is the mean average of 20, 60, and 70, that is, $50 = \frac{20+60+70}{3}$. We can thus place different amount of masses on each side of the fulcrum at the proper places, without the distribution having to be symmetrical. This will help students extend their idea of balance beyond simple symmetry with respect to the fulcrum. This in turn can help students develop as sense of balance in a set of data that goes beyond the most common and simple patterns found in students. Mokros and Russell (1991) found that, in both students and adults, ideas of mathematical balance are not fully developed beyond symmetry.

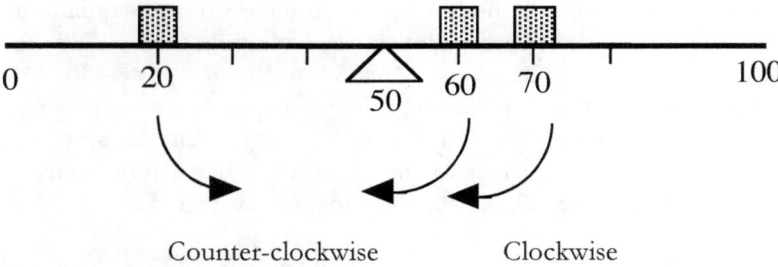

Counter-clockwise   Clockwise

Figure 13. Torques cancel out

When we have more than one unit mass on the same number, we have to weigh the sum accordingly. The system is in equilibrium because $(20 - 50) + 3(60 - 50) = 0$. Notice that 50 is the mean average of 20, 60, 60, and 60,

$50 = \frac{20+60+60+60}{4}$. This is an example of a weighted average. There are only two different values, but one of them is repeated three times. The mean will be closer to the value that is repeated.

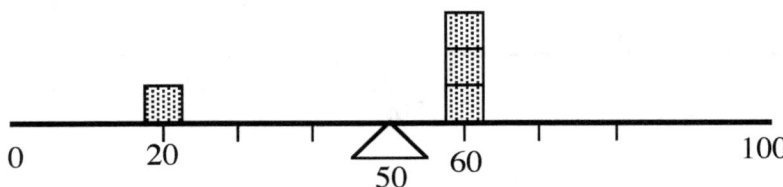

Figure 14. Unequal weights balanced

The following lever has a distribution of unit weights on the following numbers, 15, 15, 20, 60, 70, 80, 90. Let students determine whether the system is in equilibrium. Does the sum 2(15 - 50) + (20 - 50) + (60 - 50) + (70 - 50) + (80 - 50) + (90 - 50) equal zero? Students should notice the double weight on 15 in the form of the coefficient of 2(15-50).

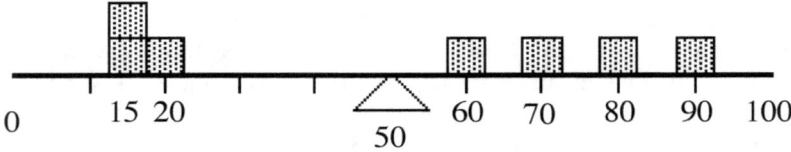

Figure 15. Is the system in equilibrium?

Let students determine whether the system in Figure 16 is in equilibrium. Is the sum of torque equal to zero? Students can compute the sum (80 – 50) + (70 – 50) + (60 – 50) + (20 – 50) + (15 – 50) = 30 + 20 + 10 – 30 – 35 = - 5. The system is not in equilibrium, it tends to rotate counter clockwise. Where do students need to add a unit weight so that the system is in equilibrium? This is equivalent to the question: What datum needs to be added to the data set 15, 20, 60, 70, 80 so that the average is 50?

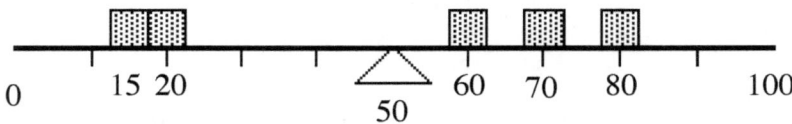

Figure 16. Is the system in equilibrium?

Let students use five unit weights to find a system in equilibrium so that four of the weights are on one side of the balance point and only one is on the other side. There are many solutions to this problem. Students can find a solution where all four weights are placed on the same number, and you can also find a solution where the weights are distributed on different spots. Students will see that to accomplish this the isolated weight needs to be placed very far from the point of balance, and that the point of balance will be very close to the other weights. In the example shown below, the point of balance is to the right of most of the weights. In the same way, an outlier can dramatically affect the mean average. The average of the numbers 35, 35, 40, 40, 100 is $\frac{35+35+40+40+100}{5}$. The average is bigger than four out of the five numbers.

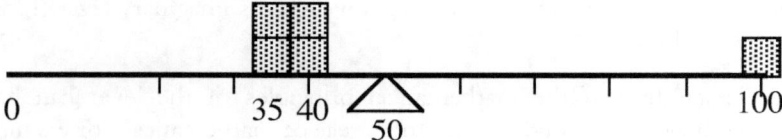

Figure 17. Effect of an outlier.

Let students distribute 5 unit weights along the meter stick so that the system is in equilibrium. Let a, b, c, d, e be the numbers students chose on the meter stick that correspond to their positions. Let them verify that the sum $(a - 50) + (b - 50) + (c - 50) + (d - 50) + (e - 50)$ is equal to zero. Let them compute the mean average of a, b, c, d, e

There is a close correlation between properties of the arithmetic average and properties of the balance point. The following properties of the average are listed by Strauss and Bichler (1988). Below each one we write the corresponding property of the balance point in a lever.

1a) The average is located between the extreme values

1b) The balance point is located between the extreme weights

2a) The sum of deviations from the average is zero

3a) The sum of moments of force with respect to the balance point is zero

3a) The average is influenced by values other than the average

3b) The balance point is changed by weights not on the balance point

4a) The average does not necessarily equal one of the values that was summed

4b) The balance point does not necessarily coincide with the place of any of the weights

5a) The average value is representative of the values that were averaged

5b) Action on the balance point is equivalent to action on the system as a whole

Once students have a firm grasp of the relation of the mean and the balance point using numbers as described above, they can tackle more general situations involving variables and using a weightless imaginary lever (Flores 1988, Flores 1995).

An understanding of the mathematical principles of the lever can have additional benefits for students in other areas of mathematics. For example, Archimedes used lever principles as a method to find the area of the parabola and the volume of the sphere (see *The Method* in his works). For further discussion of the law of the lever and its uses in other mathematical topics see Schiffer and Bowden (1984).

References

Archimedes. (2002). *The works of Archimedes*. New York: Dover.

Flores, A. (1995). Connections in proportional reasoning: Levers, arithmetic means, mixtures, batting averages, and speeds. *School Science and Mathematics, 95*, 423-430.

Flores Peñafiel, A. (1988). Giving physical sense to the average. Comunicaciones del CIMAT.

Flores Peñafiel, A. & White, A. L. (1989). Exploration of the mean as a balance point. *School Science and Mathematics, 89*, 251-257.

Hardiman, P. T., Well, A. D., & Pollatsek, A. (1984). Usefulness of a balance model. *Journal of Educational Psychology, 76*, 793-801.

Mokros, J. R., & Russell, S. J. (1991). Toward and understanding of meaning as "balance point". In R. G. Underhill (Ed.), *Proceedings of the Thirteenth Annual Meeting of the North American Chapter of the International Group for the Psychology in Mathematics Education* (Vol. 2, pp. 189-195).

Blacksburg, VA: North American Chapter of the International Group for the Psychology in Mathematics Education.

Mokros, J., & Russell, S. J. (1995). Children's concepts of average and representativeness. *Journal for Research in Mathematics Education, 26*(1), 20-39.

Pólya, George. (1977). *Mathematical Methods in Science*. Washington, DC: Mathematical Association of America, 1977.

Schiffer, M. M. and Bowden, L. (1984). *The role of mathematics in science*. Washington, DC: Mathematical Association of America.

Strauss, S., & Bichler, E. (1988). The development of children's concepts of the arithmetic average. *Journal for Research in Mathematics Education, 19*(1), 64-80.

23 LEFT AND RIGHT WITH A GRAPHING CALCULATOR[23]

This activity addresses the Data Analysis and Probability Standard for grades 9-12 of the National Council of Teachers of Mathematics (NCTM 2000), by helping students learn to select and use appropriate statistical methods to analyze data. Students work with bivariate data, display a scatterplot, describe its shape, and determine lines of best fit using technological tools.

In a mathematics methods course, each one of the future secondary teachers was asked to conduct a session in class in the same way they would conduct it in a high school. Their fellow students in turn participated in the activity as if they were high school students. We will refer to the participants as students, and to the person conducting the activity as teacher.

Students were intrigued when the teacher asked them to write the letters of the alphabet at the beginning of her high school mathematics lesson. "Hold your pen in your right hand. You will write the letters of the alphabet continuously for 30 seconds. If you get to the end of the alphabet, start

---

[23] Flores, A. and Gustafson, E. (2008). Left and right with a graphing calculator. *Ohio Journal of School Mathematics*, No. 58, 21-24. Reprinted by permission.

from the beginning again." When the teacher gave the indication to start, students wrote the letters. For most of them it was an easy task, but a couple of students struggled to write the letters. After 30 seconds the teacher asked them to stop. Then the teacher asked students to repeat the activity with their left hand. Now it was the majority of the students who struggled to write the letters, and for a couple of students it was a breeze. Student then counted how many letters they had written with each hand and wrote down the numbers. The teacher collected the data for the number of letters written in 30 seconds with each hand from each student in the class, wrote the numbers at the board (see Table 1) and then entered the two lists into the graphing calculator. Table 1 displays the data for the ten participants.

Table 1

Number of letters written with each hand

x	number of letters right hand	74	18	52	52	13	32	55	56	54	45
y	number of letters left hand	26	59	23	22	56	75	26	26	30	18

The teacher adjusted the window in the calculator to display the data as a scatter plot (Figure 1). The x-values are the number of letters written with the right hand, the y-values the number of letters written with the left hand. The teacher projected the graph so that every student could see it.

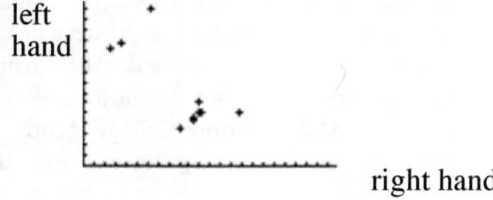

Figure 1. Scatter plot of letters written with each hand

Analyzing the graph.

The teacher led the discussion about the graph. Students drew the following conclusions. They noticed that points seem to cluster in two groups. Adding the graph of the line $y = x$, students could see this line clearly separates the two groups (Figure 2). Points above the line represent people who wrote more letters with their left hand than with their right hand, presumably, left-handed people. Points below the line represent right-handed people. Points very close to the line $y = x$ would represent people who are ambidextrous (Lanius 1999).

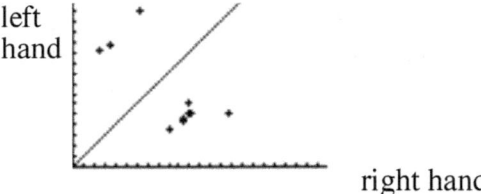

Figure 2. $y = x$ divides the two groups

You may collect data in your own classroom (or you may use the data presented here). Enter into lists L1 and L2 the number of letters written with the right hand and left hand respectively for the students in your class (here the examples are given for the data from Table 1). The window in the calculator can be set to the following values Xmin = 0, Xmax = 119, Xscl = 5, Ymin=0, Ymax = 79, Yscl = 5, Xres=1. With this window, the graph of $y = x$ will form a 45° angle.

Line of best fit

There are two methods commonly used to fit a line to a scatterplot. One is the median-median regression method; the other method is finding a line of regression using the least-squares fit. The median-median method is a simple way to introduce the idea of fitting a line to data. It is also a method that is robust to outliers (Walters, Morrell and Auer 2006). The least-squares line minimizes the sum of the squares of the distances of the data to the line. In both cases it is easy to plot a line of best fit using a graphing calculator. The instructions are for a TI-84 calculator using the median-median method. To use the median-median method press **STAT** and move the cursor to the right to **CALC**. Choose **3:Med-Med** The command Med-Med will appear on the screen. Enter the two lists that contain the data **Med-Med L1, L2**

Before pressing ENTER indicate to the calculator where to "paste" the equation in the Y= list by doing the following. Press VARS. Move the cursor to the right to **Y-VARS**. Choose **1:Function...** and choose Y2. The command now reads

Med-Med L1, L2, Y2

Press ENTER. The calculator will display the equation of the line of best fit. The equation will also appear in Y2 in the Y= list. The equation (rounded to 2 places after the decimal point) for the line of best fit using the median-median method is

$$y = -.87x + 72.31$$

When you press GRAPH, you will see a line connecting the two clusters.

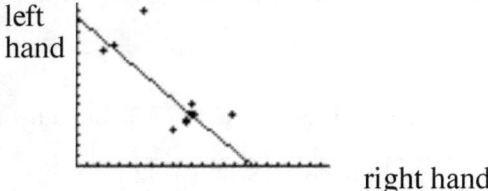

Figure 3. Med-med line of best fit for the whole group.

Instructions for using the least square method are equally simple (you would choose LinReg(ax+b) instead of Med-Med). For the data in this activity, the equation for the least squares linear regression function is very similar to the one obtained above (again numbers have been rounded), $y = -.79 x + 71.9$. The graph of the least squares line (Figure 4) looks also very similar to the previous one.

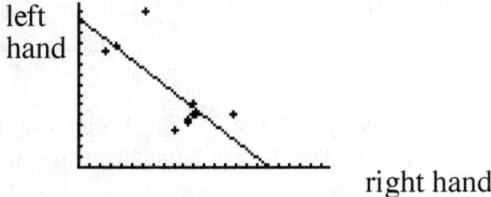

Figure 4. Least squares line of best fit.

Discussion of the line of best fit for all the data.

The line of best fit for all the data has a negative slope. The interpretation would be that the more letters people write with their right hand, the fewer letters they write with their left hand. This line of fit reflects the different behavior of the two clusters. Right-handed people write more letters with their right hand than with their left hand, but for left-handed people the reverse is true. So people who wrote very few letters with their right hand, wrote many letters with their left hand.

However, because we have two clearly distinct subgroups, this is a case where trying to describe globally the data can be misleading. If we look at each cluster separately, we see that within each subgroup, people who write more letters with their right hand than other people in the same group tend also to write more letters with their left hand.

Line of best fit for each subgroup.

To study the two subgroups separately, we generate two sub-lists, one containing only the data for left-handed people (Table 2) and one for right-handed people (Table 3). We then generate the line of best fit for each table (Figures 5 and 6).

Table 2
Number of letters written with each hand – left-handed people

x	number of letters right hand	18	13	32
y	number of letters left hand	59	56	75

Table 3
Number of letters written with each hand— right handed people

X	number of letters right hand	74	52	52	55	56	54	45
Y	number of letters left hand	26	23	22	26	26	30	18

Line for right-handed people $y = .38\,x + 7.74$. Line for left handed people $y = x + 42.33$. Given the small numbers in each group, it is probably not very significant to focus on the difference in slope values for the graphs of left handed people and right handed people (.38 vs. 1). What is striking is that for each subgroup the slope is positive, whereas for the group as a whole the slope is negative.

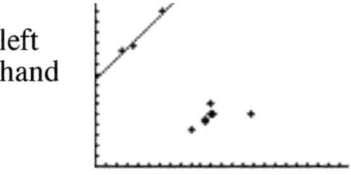

left
hand

right hand

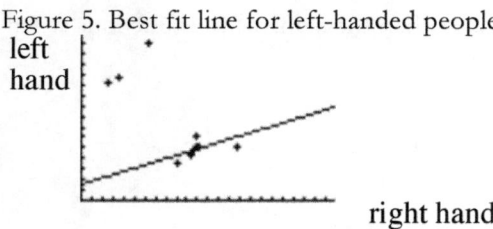

Figure 5. Best fit line for left-handed people.

Figure 6. Best fit line for right handed people.

Dominant hand vs. non-dominant hand.

A better way to think about comparing the speed of different people when writing letters would be to think in terms of dominant hand, rather than left-handed or right-handed. We can arrange the data by dominant and non-dominant hand instead (Table 4), and obtain the corresponding scatter plot (Figure 7).

Table 4
Number of letters written with dominant and non-dominant hand

x	number letters dominant hand	74	59	52	52	56	75	55	56	54	45
y	number letters non-dominant hand	26	18	23	22	13	32	26	26	30	18

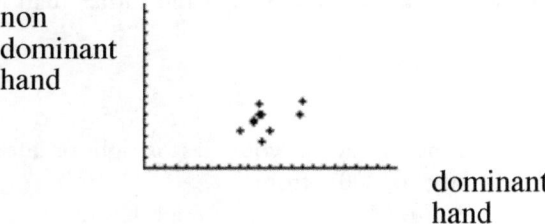

Figure 7. Letters with dominant hand vs. non-dominant hand.

The equation of the med-med line of best fit (with rounded numbers) is $y = .18\ x + 13.67$. The corresponding graph is shown in Figure 8. The interpretation is that in general, people who are faster than others writing letters with their dominant hand will also tend to be faster to write letters with their non-dominant hand than others. Of course, there are individual cases in the graph where that is not the case. Can you find them? Notice also that the slope for the group as a whole, is now positive (0.18).

However, it is smaller than the slope of the functions when the two groups were graphed separately.

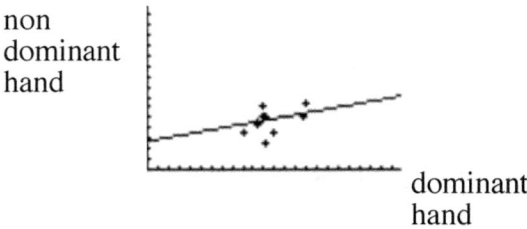

non
dominant
hand

dominant
hand

Figure 8. Best fit for dominant vs. non-dominant hand.

Concluding remarks

The example presented here is an instance where the global trend masks an opposite trend in each of the subgroups. This not an uncommon situation. For example, an airline can have a better on-time record than another airline for each of the airports where both fly, but have an overall worse on-time record, due to the fact that the first airline flies more often to airports where there is likely to be delays due to bad weather, and the second airlines flies more often to cities where the weather is sunny most of the year. The ability to look beyond a global trend and focus on trends in subgroups is an important one. By working on examples like the one presented here, students can also learn to re-conceptualize the problem and find ways to describe a situation that reveal better the relation between the variables, as they did by thinking in terms of dominant hand rather than left- or right-handedness.

References

Lanius, C. (1999). What percentage of your class is right or left handed? Retrieved February 26, 2008, from http://math.rice.edu/~lanius/Algebra/rightleft.html

National Council of Teachers of Mathematics. *Principles and Standards for School Mathematics*. Reston, VA: National Council of Teachers of Mathematics, 2000.

Walters, Elizabeth J. Christopher H. Morrell, and Richard E. Auer. "An Investigation of the Median-Median Method of Linear Regression" *Journal of Statistics Education* Volume 14, Number 2 (2006). Retrieved August 7, 2008 from www.amstat.org/publications/jse/v14n2/morrell.html

24 SUBTRACTION OF POSITIVE AND NEGATIVE NUMBERS: THE DIFFERENCE AND COMPLETION APPROACHES WITH CHIPS[24]

Diverse contexts such as *take away*, *comparison*, and *completion* give rise to subtraction problems. The take away interpretation of subtraction has been used with of chips of two colors to help students understand addition and subtraction of integers (Bennet and Musser 1981). In this article we will illustrate how the difference and completion interpretations of subtraction may be used with chips of two colors to develop further understanding of subtraction of negative and positive numbers. We will use black chips to represent positive numbers and white chips to represent negative numbers. When using chips of different colors for addition and subtraction of integers, the basic principle is that chips of different colors cancel each other when joined together. Each of the collections in Figure 1 represents zero.

Figure 1. Three representations of 0.

We can also represent any other integer in multiple ways, by adding zero-pairs as needed. The number 2 can be represented by two black chips, but also by a collection of three black chips and one white chip.

---

[24] Flores, A. (2008). Subtraction of positive and negative numbers: The difference and completion approaches with chips. *Mathematics Teaching in the Middle School*, *14*(1), 21-23. Copyright National Council of Teachers of Mathematics. Used by permission.

Figure 2. Three representations of 2.

Comparing collections: The difference

Finding the difference is useful in situations in which we compare collections. To find the difference 7 – 3 we match as many objects from the first collection as we can to the objects in the second collection. The first 3 objects in the collection of 7 are matched with the objects in the second collection. The number of unmatched objects is the difference, 4 in this case.

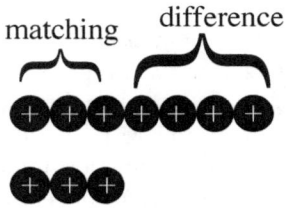

Figure 3. Matched part and difference.

The same approach works to compare negative numbers.

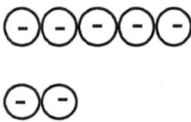

Figure 4. Finding -5 - (-2) = -3

What if the first collection has fewer objects than the second, for example in 2 - 5?

Figure 5. Not enough to match the second collection.

Because we do not have enough objects in the collection of 2 objects to match all the objects in the second collection of 5 objects, we need to represent 2 in a different way by adding three zero-pairs.

Figure 6. Another representation for 2

Now we can match five objects in the collection representing 2 with all objects in the collection representing 5. The difference is -3.

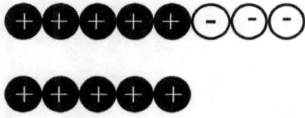

Figure 7. Finding the difference 2 - 5 = -3

We can use the same method of adding zeros when comparing collections of chips of different color.

Figure 8. Comparing 5 and (-1)

To find the difference 5 - (-1) by matching, we need to add one negative chip and a positive chip in the representation of 5. We can now match the white chips and the six black chips will be the difference.

Figure 9. Finding the difference 5 - (-1) = 6

To find the difference -3 – 4 we encounter the same problem of not being able to match because the chips are of different color.

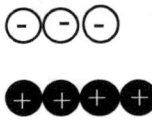

Figure 10. Trying to compare -3 and 4

We also need to add zero-pairs. By adding four zero-pairs we will have enough black chips in the representation of -3 to match all the black chips in the collection of 4.

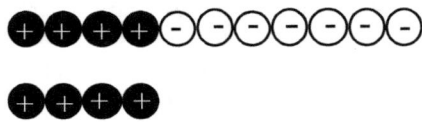

Figure 11. Finding the difference -3 - 4 = -7

Completion

Another interpretation of subtraction is based on completion or finding the missing addend. Here the idea is to find what number do we need to add to a given number to get another number.

In this approach, to subtract 5 - 2, we need to find the number that added to 2 gives 5, that is, 2 + ? = 5. Because 2 + 3 = 5, we have 5 - 2 = 3. We can illustrate this subtraction with chips as a missing addend problem.

start | missing addend

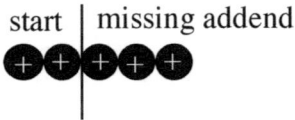

target number: 5

Figure 12. 5 - 2 = 3

For the subtraction 2 - 5, we start with 5 black chips and our target is a collection whose value is 2. We need to find the missing addend, 5 + ? = 2. We need to add 3 negative chips to cancel the excess in 5 to obtain 2.

start | missing addend

target number: 2

Figure 13. 2 - 5 = -3

To find 3 - (-1), starting with (-1) we need to add 4 positive chips to end up with a collection whose value is 3.

start | missing addend

target number: 3

Figure 14. 3 − (-1) = 4

For the subtraction -3 - (-4), starting with (-4) we need to add one positive chip to end up with -3.

start | missing addend

target number: -3

Figure 15. -3 − (-4) = 1

Another interpretation: Distance

A student found that she could solve a problem that involved subtracting a negative number from a positive number, like 8 - (-5), by taking the corresponding number of chips of each color and then count the total number of chips.

Figure 16. 8 - (-5) = 13

This method is related to finding the distance between numbers on the number line using arrows. The difference $a - b$ is the distance from a to b (taking direction into account). To solve $8 − (-5)$, we need go from -5 to 8.

We can see that we need to walk 13 units in the positive direction.

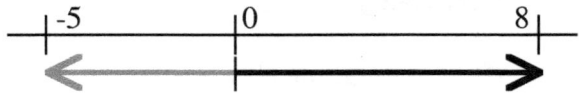

Figure 17. Directed arrows to illustrate 8 - (-5) = 13

Concluding remarks

When working with chips to represent subtraction problems, students need to pay attention to the very different roles that the numbers take, which reflects the fact that subtraction is not commutative. 5 – 3 and 3 – 5 give very different results. The role that 3 has in each of these problems is also very different. Students need to think clearly about the different roles of the first and second numbers in a subtraction problem, both in the difference and the completion contexts.

By using a concrete approach first, students will internalize understanding, and make sense of the rules to subtract integers. As students understand these rules, they will need to rely less and less on the concrete models, and eventually chips will not be necessary.

Reference

Bennett, Albert B., and Gary L. Musser. "A Concrete Approach to Integer Addition and Subtraction." In *Activities for Junior High School and Middle School Mathematics*, edited by Kenneth E. Easterday, Loren L. Henry and F. Morgan Simpson, 24-27, 172. Reston, VA: National Council of Teachers of Mathematics, 1981.

25 THE PYTHAGOREAN THEOREM WITH JELLY BEANS[25]

An empirical exploratory approach to geometry in the middle grades is advocated in *Principles and Standards for School Mathematics* (NCTM 2000). Students can explore and examine a variety of geometric shapes and discover their characteristics and properties using hands-on materials. Students can also create inductive arguments about the Pythagorean relationship. This empirical approach to the theorem will lay the foundation for later analytical proofs. Incorporating experimentation as part of middle school geometry is also consistent with research on how students learn best. According to Van Hiele, students need opportunities to develop their geometrical thinking through five levels (Van Hiele 1986; Fuys, Geddes, and Tischler 1988). Especially relevant in the middle school are the second level that corresponds to *Analysis of properties* and the third level of *Informal deduction*. At the analysis of properties level a student can go beyond global perception of shapes and analyze figures in terms of their components, describe their parts and list their properties. Descriptions are used instead of definitions. Students discover and prove properties or rules empirically (for example by folding, measuring, or using a grid or a diagram). At the informal deduction level a student can understand the role of definitions, the

---

[25] Yun, J. O. and Flores, A. (2008). The Pythagorean theorem with jelly beans. *Mathematics Teaching in the Middle School, 14*(4), 202-207. Copyright National Council of Teachers of Mathematics. Reprinted by permission.

relationships between figures; can order figures hierarchically according to their characteristics; can deduce facts logically from previously accepted facts using informal arguments. The activities presented here are appropriate to help students make the transition from the level of development of geometrical thinking that corresponds to the level of analysis of properties and empirical verification to the next level of informal deduction. The first two activities presented here allow students in the middle grades to explore the Pythagorean theorem and one extension by first using jelly beans to measure the areas of squares and semicircles. In the third activity students use the result they obtained for squares, and use deductive thinking to establish the result for semicircles on the sides of the right triangle by applying equivalencies of algebraic expressions. In the fourth activity, students use their discovery about the areas of semicircles and deductive thinking to establish further relations among areas.

The tools needed for the first activity are a cardboard mat that is formed by a right triangle and the corresponding squares on its sides (see Figure 1), a cardboard fence that surrounds the three squares (Figure 2), another fence around the central triangle (Figure 3), and enough jelly beans to cover the two smaller squares with one layer.

There are several ways in which the activities could be conducted. The teacher can do the first activity as a demonstration and lead the discussion to make the relationship between the areas of the squares explicit. Students can also work in small groups of four around a mat. Different groups could have different mats based on different shaped right triangles, such as isosceles right triangles (Figure 3) and scalene right triangles (Figure 4) and then compare results. Or students could move from one station to the next where the layout of the mat is different and conduct a similar experiment.

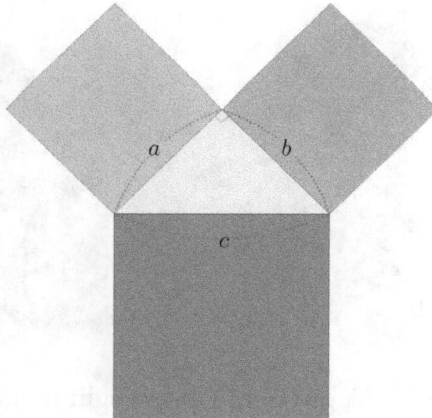

Figure 1. The mat.

Figure 2. The mat with the outside fence

Figure 3. The triangular fence is added.

Figure 4. A different shaped right triangle.

To construct the mat use a drawing or geometry computer program or the drawing tools included with word processors, print the colored figures, and paste them on cardboard about the size of a shoebox. To make the fences cut strips about 2 cm wide of corrugated cardboard, with the cuts perpendicular to the corrugates so that the strips can be bent into the desired form. Paste the strips on the mat along the outside figures with instant glue. Make the fence for the inner right triangle separately. Jelly beans work better than black beans or grains of rice. With jelly beans it is easier to make just one layer, and guarantee that any empty spaces are very small compared to the jelly beans. In addition, students prefer the colorful jelly beans rather than the black beans. Of course, students need to abstain from eating any jelly beans they use to cover the shapes.

Although jelly beans are three-dimensional objects, by using only one layer, we can use them to measure area. By using one layer only of jelly beans we are essentially using the cross section of the jelly beans, which is two-dimensional. In this activity students do not have to count the number of jelly beans to compare areas; they use only the global amount. As they move jelly beans from one section of the mat to another, they will see whether all the jelly beans fit into the new shape or not.

Activity 1. The Pythagorean relationship.

For this activity, students will use the mat that consists of a right triangle with squares on each of the sides of the triangle. With the cardboard fence glued around the three squares (Figure 2a), students insert the triangular fence around the triangle (Figure 2b). They should notice that by doing so, three squares are formed on the sides of the right triangle. Then they pour jelly beans in the two squares on the legs of the right triangle and make sure that those jelly beans cover completely the two squares, without leaving gaps, and forming just one layer (Figure 5a). Students remove the cardboard triangle that is inside the frame (Figure 5b), and incline the bottom cardboard to let those jelly beans slide into the square on the hypotenuse (Figure 5c). Finally, they insert the triangle fence in its original place again (Figure 5d) and flatten out the jelly beans. Students verify whether the jelly beans fit in and completely cover the square on the hypotenuse with one layer (Figure 5e). The teacher guides students to make explicit and express in their own words what they found about the sum of the areas of the squares on the legs of the right triangle compared to the area of the square on the hypotenuse. Then students can label the two legs of the right triangle as a and b, and the hypotenuse as c, and express the relationship using an algebraic equation, $a^2 + b^2 = c^2$.

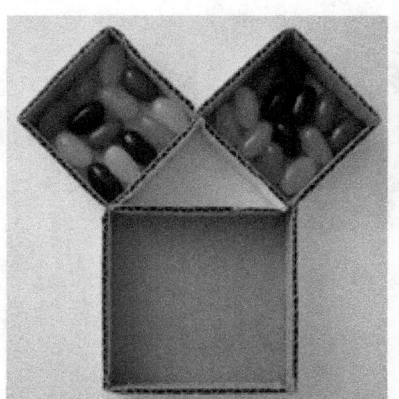 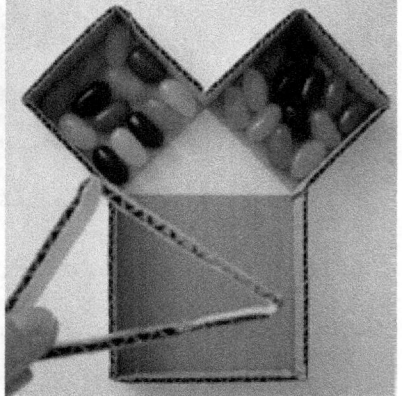

Figure 5a. Two squares on the legs are covered.   Figure 5b.
Removing the inside fence.

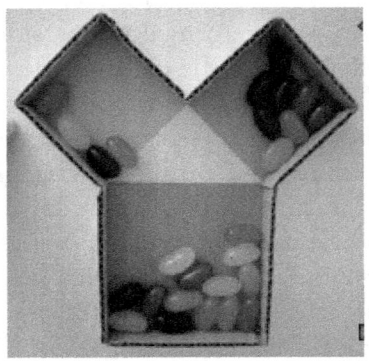 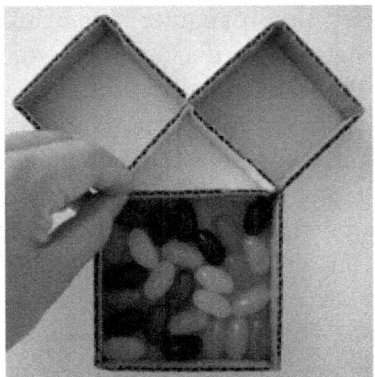

Figure 5c. Slide the jelly beans. Figure 5d. Reinsert the triangle fence.

Figure 5e. Jelly beans cover the square on the hypotenuse.

Activity 2. Extension of the Pythagorean theorem.

Students can also use jelly beans to explore the relationships between areas when similar shapes are constructed on the sides of a right triangle. Students use a mat with three semicircles around a right triangle (Figure 6). The diameters of the semicircles are congruent to the corresponding sides of the right triangle. With the cardboard fence around the three semicircles glued to the mat, students insert the triangular fence around the right triangle. They need to notice that the two fences together form three semicircles on the sides of the right triangle (Figure 7). In the same way as in the first activity, students pour jelly beans in the two semicircles on the legs of the right triangle and make sure that those jelly beans cover completely the two semicircles, without leaving gaps, and forming just one layer (Figure 8). Then they remove the cardboard triangle that is inside the frame, incline the bottom cardboard to let those jelly beans slide into the semicircle on the hypotenuse. Finally they insert the triangle fence in its original place again, and flatten out the jelly beans and verify whether the jelly beans fit in and completely cover the semicircle on the hypotenuse with one layer. Students can describe the relationship they see between the sum of the areas of the semicircles on the legs of the right triangle and the area of the semicircle on the hypotenuse in their own words.

The teacher will also guide students to see that in each of the activities the three shapes on the sides of the right triangle are similar to each other, that is, squares are similar to other squares, and semicircles are similar to other semicircles. These two activities correspond to the second Van Hiele level because the verifications are done empirically.

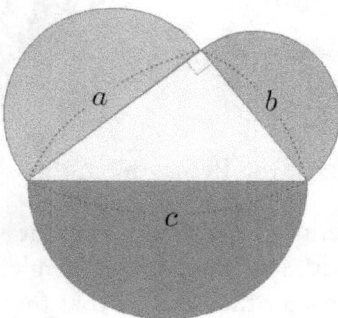

Figure 6. Three semicircles around a right triangle.

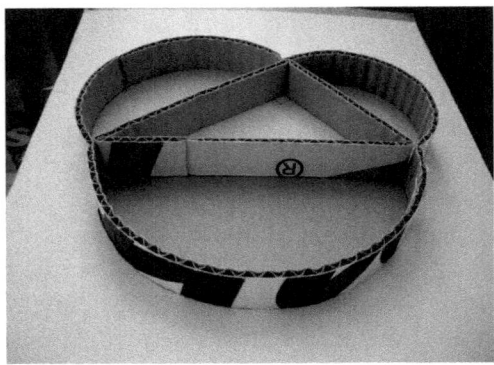

Figure 7. The fences form three semicircles.

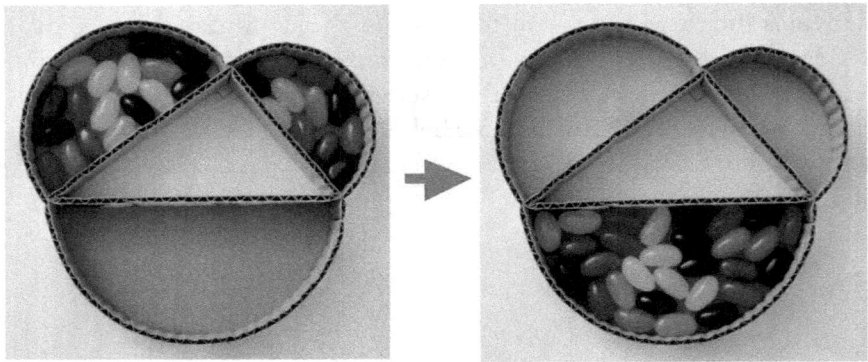

Figure 8. Extension of the Pythagorean theorem for semicircles.

Other extensions. Students can also experiment with different shapes constructed on the sides of the right triangle, as long as all three shapes are similar to each other and so that their corresponding sides are placed on the sides of the right triangle (see Figure 9). Of course these extensions to the Pythagorean theorem in terms of similar shapes on the sides of a right triangle are not new (Euclid Book 6, Prop. 31; Pólya 1948; Flores Peñafiel 1992), but usually students are surprised that the Pythagorean relationship holds also for shapes other than squares.

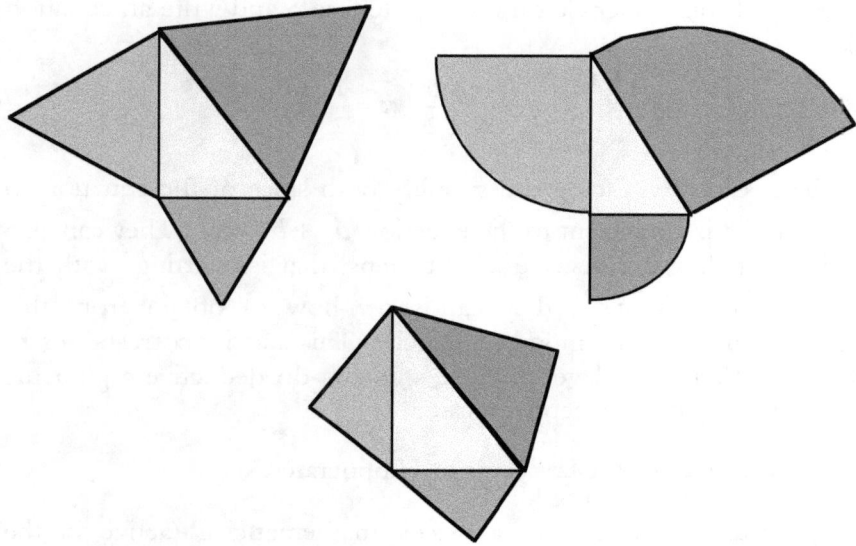

Figure 9. Other extensions of the Pythagorean theorem

Activity 3. Formulas for the areas of the semicircles and connections with algebra.

Students can derive the extension of the Pythagorean theorem with semicircles by using some algebraic skills such as factoring and simplifying algebraic expressions. They can do this activity working in pairs or small groups. Students can denote the two legs of the right triangle as a and b, and the hypotenuse as c (Figure 6), and express the relationship among the areas of the squares on the sides of the right triangle as $a^2 + b^2 = c^2$. They need also remember that the diameter of each semicircle is congruent to the corresponding side of the triangle, and that the sum of the areas of the semicircles on the legs of the right triangle was equal to the area of the semicircle on the hypotenuse. The teacher can guide students so they can write algebraic expressions for the areas of the semicircles, and for their relationships. Students can write an algebraic expression for each radius of the semicircles in terms of the length of each side of the triangle, and write and expression for the area of each semicircle. Students can use algebraic notation to state that the sum of the areas of the semicircles on the legs a and b of the right triangle is equal to the area of the semicircle on the hypotenuse c, and write an equation like

$$\frac{1}{2} \times \left(\frac{a}{2}\right)^2 \pi + \frac{1}{2} \times \left(\frac{b}{2}\right)^2 \pi = \frac{1}{2} \times \left(\frac{c}{2}\right)^2 \pi$$

Students can then factor and simplify both sides of the equation to show that it is equivalent to the equation $a^2 + b^2 = c^2$. They can also verify that they can reverse all the steps, that is, starting with the equation $a^2 + b^2 = c^2$, they can show how to obtain from this equation the relation among semicircles. This activity corresponds to the third Van Hiele level because students do deductive arguments based on previously accepted facts.

Activity 4. Area of the crescents of Hippocrates.

Hippocrates of Chios was a Greek mathematician (active in the second half of the fifth century BC), who made the remarkable discovery of two figures bounded by arcs of circles that have the same area as a figure bounded by straight lines. In Figure 10, semicircles are constructed on the sides of the right triangle ABC. However, the semicircle on the hypotenuse, rather than being on the outside of the triangle as in Activity 2, is drawn on the inside. Students who have difficulty seeing this semicircle could look at three semicircles on the outside of a right triangle as in Figure 6 and then think of reflecting the largest semicircle across the hypotenuse. The semicircle on the hypotenuse overlaps with the triangle (yellow), and also overlaps partially with the semicircles on the legs (green). The part of the semicircle on each leg that is completely outside the semicircle on the hypotenuse makes a moon crescent (blue). Students can express the area of the semicircle on the hypotenuse as $t + a_1 + a_2$ (Figure 10). The areas of the semicircles on the legs are $c_1 + a_1$ and $c_2 + a_2$. Using their discovery that the area of the semicircle on the hypotenuse is equal to the sum of the areas of the semicircles on the legs of the right triangle, and the expression of areas stated above, students can find a relation between the sum of the areas of the two crescents and the area of the right triangle ABC. This activity corresponds also to the level of informal deduction.

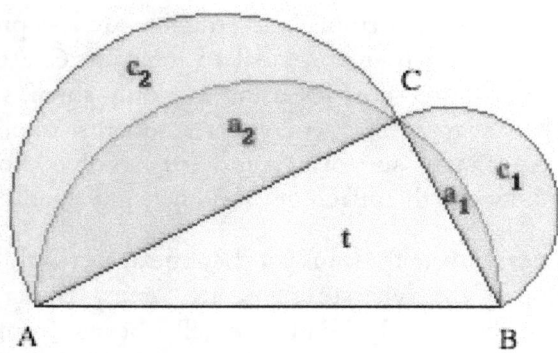

Figure 10. Crescents formed by semicircles.

Concluding remarks

Many students dislike mathematics when they have to memorize things that they cannot verify and understand on their own. Often, when the Pythagorean theorem is first introduced using a deductive approach, students find it hard to follow the steps of the proofs. And sometimes, even when they manage to follow each step to reach the conclusion, many do not seem to understand what it all means. An approach that proves the Pythagorean theorem using only axioms or previously demonstrated theorems corresponds to the fourth level of development of geometric thinking, axiomatic deduction. The difficulties students experience in a deductive presentation are because the level that the teacher is operating (fourth level) does not correspond to the levels most students in middle school operate (second and third level).

In the case of the Pythagorean theorem, sometimes students are confused about what areas they need to add to get which area, and what are the segments that are being compared. Although nowadays there are amazing applet programs that students can use to visualize the relationship, some students find that it is an experience that asks them to believe based on an outside authority because all they can do is watch untouchable moving clips, rather than experiment directly with more concrete objects. The jelly beans activities are helpful for students to understand the Pythagorean theorem conceptually, rather than just memorizing a formula, and students are less prone to forget what areas added together are equal to what other area. When the teacher prepares more activities using not only squares on the sides but also other shapes like semicircles, students remember better. They find the results surprising, and get involved in establishing connections between the different cases by using algebraic notation.

These activities with jelly beans are beneficial not only for students. Prospective and in-service teachers also enjoy doing the activity, removing the fence and sliding the jelly beans from the smaller shapes on the legs of the triangle to the bigger shape on the hypotenuse and verifying that indeed the shape gets covered (Yun 2007).

References

Euclid. *The Thirteen Books of Euclid's Elements.* Translated by Thomas L. Heath. Vol. 2. New York: Dover, 1956.

Flores Peñafiel, Alfinio. "La Feria de Pitágoras (2a parte)." *Educación Matemática,* 4(2) (1992): 62-78.

Fuys, David, Dorothy Geddes, and Rosamond Tischler. *The Van Hiele Model of Thinking in Geometry among Adolescents.* Reston, Va.: National Council of Teachers of Mathematics, 1988.

National Council of Teachers of Mathematics. *Principles and Standards for School Mathematics.* Reston, Va.: National Council of Teachers of Mathematics, 2000.

Pólya, George. "Generalization, Specialization, Analogy." *American Mathematical Monthly,* 55(4)(1948): 241-243.

Van Hiele, Pierre M. *Structure and Insight: A Theory of Mathematics Education.* Orlando, Fla.: Academic Press, 1986.

Yun, Jeong Oak. (2007). "Three activities for teaching geometry 5 - 8." Unpublished manuscript. Tempe, Ariz.: Arizona State University.

26 EXPLORATIONS WITH 142857: CONNECTING THE ELEMENTARY WITH THE ADVANCED[26]

This article presents an interesting activity for students and some further explorations that connect some of the abstract mathematical concepts learned in college level courses with the mathematics taught in schools.

Explorations for students

When students multiply 142857 by each of the numbers 1, 2, 3, 4, 5, 6 they are pleasantly surprised to observe that the same digits 1, 4, 2, 8, 5, 7 appear in each of the results. Some will notice that the digits also appear in the same "order" but shifted in a cyclic way. For example, for $142857 \times 2 = 285714$, the digit 7 comes after 5 in both cases. They also notice that when a digit "runs out of space," it goes to the end of the sequence when shifting to the left, or at the beginning when shifting to the right. 142857 is an example of a cyclic number. For more tricks with cyclic numbers see the chapter "Cyclic numbers" in Gardner (1979).

$$142857 \times 1 = 142857$$

$$142857 \times 2 = 285714$$

[26] Flores, A. (2008). Explorations with 142857: Connecting the elementary with the advanced. *Mathematics Teacher*, *101*(6), 18-21. Copyright National Council of Teachers of Mathematics. Reprinted by permission.

$$142857 \times 3 = 428571$$

$$142857 \times 4 = 571428$$

$$142857 \times 5 = 714285$$

$$142857 \times 6 = 857142$$

With encouragement to look for patterns, students can predict how the digits will appear above without having to do all the multiplications, by simply placing the leading digit in ascending order, and then accommodating the other digits in a cyclic way.

Students can try to find out why the same digits appear in all the products, and why the order is always the same, except for a cyclic shift. Some will wonder what happens when multiplying 142857 by 7. To their surprise they get 999999. From here they can see that .999999 ÷ 7 = .142847, and this may suggest looking at the decimal expansion of 1/7 = .142857142857.... The digits of 142857 form the repeating decimal expansion of 1/7. If students observe the decimal expansions of the different fractions 2/7, 3/7, and so on, they will notice the relation to the digits obtained when multiplying 142857 by 2, 3 and so on.

Further explorations using long division

When we divide 1 ÷ 7 using the common long division algorithm, we see that all the possible remainders 1, 2, 3, 4, 5, 6, appear in the process, and then they will repeat themselves in the same order: 1, 3, 2, 6, 4, 5. At this point we suggest that students look at a shortened version of the algorithm with just the remainders written down (figure 1). Once the remainder 1 appears, the process is going to repeat itself, giving rise again to the digits 142857 in the decimal expansion of the quotient. If we take any number between 1 and 6, and divide it by 7, we will essentially be starting the above process at a different place in the remainder sequence. For example, for 6 ÷ 7, the sequence of digits in the quotient would start with 8 (figure 2). Therefore, the digits of the new quotient will be the same as the digits in 1 ÷ 7, only starting at a different place.

$$0.1428571\ldots$$
$$7\overline{)1.0000000\ldots}$$

3

2

6

4

5

1

$$0.8571\ldots$$
$$7\overline{)6.0000\ldots}$$

4

5

1

3

Figure 1. Remainders in $1 \div 7$ Figure 2. Remainders in $6 \div 7$

More advanced explorations

Notice that multiplying by 2 shifts the digits two places, and multiplying by 4 the digits shift four places, but this pattern does not work for the other numbers; multiplying by 3 shifted the digits only one place. Is there a pattern or a relation between what numbers were multiplied by 142857 and how many places the digits are shifted? The number of places the digits in 142857 get shifted can be 0, 1, 2, 3, 4, 5. The digits shift in a cyclic way, so we can represent how many places a given digit shifts using a "clock" with six hours (0, 1, 2, 3, 4, 5) (Figure 3). Here shifting 3 places and then shifting 4 more is the same as shifting just one place. That is, for 3 + 4, the remainder after dividing by 6 will be 1 or $3 + 4 \equiv 1 \pmod 6$. This notation is read as "congruent modulo six." The remainders for dividing by 6, {0, 1, 2, 3, 4, 5}, form an additive group when the numbers are added in the usual way and reduced modulo 6. For example $3 + 5 \equiv 2 \pmod 6$, and $4 + 2 \equiv 0 \pmod 6$. We can verify that the additive identity is 0, the operation is closed and commutative, and each element has an additive inverse, and therefore this forms a group. It is also clear that the number 1 generates the whole group.

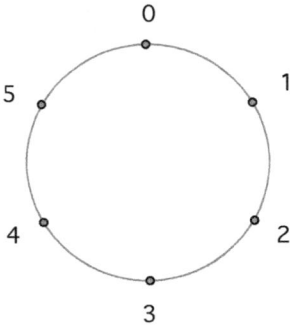

Figure 3. Additive group modulo 6.

A useful hint to understand how many places the digits get shifted when multiplying by a given number is to look at the order in which the remainders 1, 3, 2, 6, 4, 5 appear in the long division procedure. We will be able to tell how many places the digits in the quotient will be shifted by looking at where the corresponding remainder appears in the sequence. Thus, when dividing $3 \div 7$ the digits will be shifted one place to the left (and 1 will go to the end), because 3 is the first new remainder to appear in the algorithm above. When dividing $2 \div 7$ the digits will be shifted 2 places, when dividing $6 \div 7$ the digits will be shifted 3 places, and so on.

A deeper look

Looking again at the remainders for $1 \div 7$, notice that each number in the sequence 1, 3, 2, 6, 4, 5 can be obtained from the one before by multiplying by 3 and reducing modulo 7. Thus $3 \times 3 = 9$, and $9 \equiv 2$ (mod 7), $3 \times 2 = 6$, and $6 \times 3 \equiv 4$ (mod 7), etc. We can write the remainders in a cyclic diagram (Figure 4) and verify that the remainders (1, 2, 3, 4, 5, 6) we obtain when dividing by 7 form a multiplicative group (modulo 7). We multiply the numbers as usual, but reduce modulo 7. So for example, $2 \times 5 \equiv 3$ (modulo 7), $6 \times 6 \equiv 1$ (mod 7). We can verify that the multiplicative identity is 1, the operation is closed, and it is commutative, and each element has a multiplicative inverse. We can notice that not all numbers generate the whole group. The number 3 generates the whole group, but 2 does not, because when multiplying by 2 we go back to 1 after only two steps, $4 \times 2 \equiv 1$ (mod 7) (figure 5). We can also write the numbers as powers of 3, modulo 7 (Figure 6). The sixth power will bring us back to 1, $3^6 \equiv 1$ (mod 7).

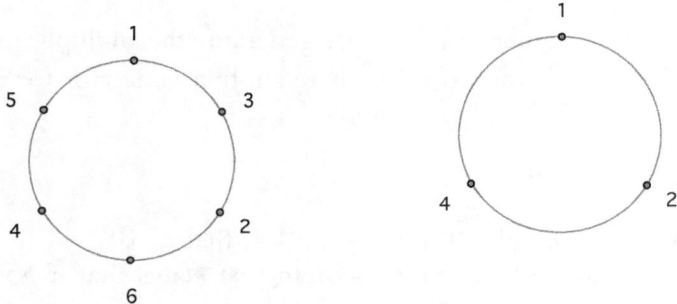

Figure 4. Multiplicative group modulo 7Figure 5. A proper subgroup.

$$3^0 \quad 3^1 \quad 3^2 \quad 3^3 \quad 3^4 \quad 3^5$$

$$\downarrow \quad \downarrow \quad \downarrow \quad \downarrow \quad \downarrow \quad \downarrow$$

$$1 \quad 3 \quad 2 \quad 6 \quad 4 \quad 5$$

Figure 6. Powers of 3 modulo 7

Why does 3 have such a prominent role in this context? We can remember that in the long division algorithm we multiply a remainder by 10 for the next step. Because $10 \equiv 3$ (mod 7), multiplying by 10 (mod 7) is the same as multiplying by 3 (modulo 7).

A one to one correspondence between the two groups {0, 1, 2, 3, 4, 5}, + (mod 6) and {1, 2, 3, 4, 5, 6}, × (mod 7) that preserves operations can be given by pairing the respective generators 1 and 3 (Figure 7).

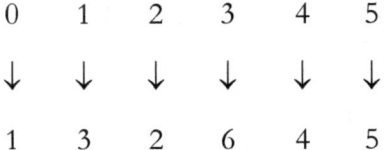

Figure 7. Isomorphism of groups

Exercise. Is there another number that generates the multiplicative group {1, 2, 3, 4, 5, 6}, × (mod 7)? Is there another generator for the additive group {0, 1, 2, 3, 4, 5}, + (mod 6)?

Another cyclic number

We saw above that 7 divides $999999 = 10^6 - 1$, that is, $10^{7-1} \equiv 1$ (mod 7). This is an instance of Fermat's theorem that states that if p is a prime that does not divide the integer a, then $a^{p-1} \equiv 1$ (mod p) (Courant and Robbins 1996). So, primes other than 2 and 5 will divide a string of 9s of length $p - 1$. The prime 7 requires the full length of six 9s. In some cases, such as 3, 11, and 13, the prime also divides a shorter string of 9s, $3 \times 3 = 9$, $11 \times 9 = 99$, $13 \times 76923 = 999999$. The next prime that requires the full length in the string of 9s is 17, $17 \times 588235294117647 = 9999999999999999$. The digits in the decimal period of $1/17 = .0588235294117647...$ give rise to another cyclic number. The order of the remainders when using the algorithm $17)\overline{1}$ will tell students how many places the digits are shifted when multiplying 058823594117647 by each of the numbers between 1 and 16 (Figure 8). The pairing preserves operations between the multiplicative group modulo 17 and the additive group modulo 16. For example, $15 \times 6 \equiv 5$ (mod 17) and $2 + 5 \equiv 7$ (mod 16).

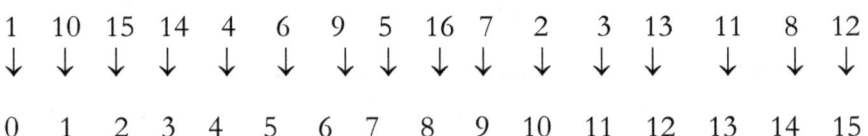

Figure 8. Another isomorphism of groups.

Concluding remarks

University mathematics courses do not always provide the opportunity to make connections between the advanced topics and the mathematics taught in the middle school or high school. Activities like the ones described in this article invite such connections. Looking at concrete or particular examples provides a better grasp of the abstract concepts. Oftentimes, looking at the mathematics they teach from a more advanced point of view can help teachers develop this deeper understanding. Klein (2004) and Rademacher (1983) give us great examples of connecting the advanced with the elementary. The more opportunities teachers have to make such connections the better.

References

Courant, Richard and Herbert Robbins. *What is Mathematics?* 2nd edition, revised by Ian Stewart. New York: Oxford University Press, 1996.

Gardner, Martin. "Cyclic Numbers." Ch. 10 in *Mathematical Circus: More Puzzles, Games, Paradoxes and Other Mathematical Entertainments from Scientific American.* New York: Knopf, pp. 111-122, 1979.

Klein, Felix. *Elementary mathematics from an advanced standpoint: arithmetic, algebra, analysis.* New York, Dover Publications, 2004

Rademacher, Hans. *Higher mathematics from an elementary point of view.* Boston: Birkhäuser, 1983.

27 THE FINGER AND THE MOON[27]

There is a Tanzanian proverb, "I pointed out to you the moon and all you saw was the tip of my finger." The proverb came to my mind as I was looking at cases of students working with decimals and the difficulties they were having (Barnett, Goldenstein, and Jackson 1994). It dawned on me that part of the difficulty is that we use the decimal point as a pointer to indicate the place value of the digits around it, but many students focus on the finger (the point) and not the moon (place value).

Students are not the only ones that sometimes focus on the finger rather than on the moon. In a lesson with teachers on ways to teach decimals in a meaningful way, we used base-ten blocks as one of our learning tools. After stating that we would use the flat as our unit, teachers promptly found the value of the other pieces. They figured out that the long would be 1/10 or .1, and the small cube would be 1/100 or .01. They agreed that the result of multiplying a long times 10 would be a flat, that is .1 times 10 would be 1, because ten longs make a flat, and multiplying the small cube by 10 would be a long, that is, .01 times 10 would give .1. I asked them to represent .13 on a place value mat, both with blocks and with digits, and to represent the decimal point with a coin or another object. Then I asked them to multiply each of the pieces on their display by 10 and represent the answer on the mat, both with blocks and with digits. Teachers promptly grabbed the corresponding bigger pieces and put them in the corresponding columns, and wrote down the digits in their corresponding columns. They saw, to their surprise, that the decimal point had not moved. All of them had

---

[27] Flores, A. (2007). The finger and the moon. *Mathematics Teaching in the Middle School, 13*(3), 132-133. Copyright National Council of Teachers of Mathematics. Reprinted by permission.

learned that to multiply a number by 10, you simply move the decimal point right or left following a set of rules.

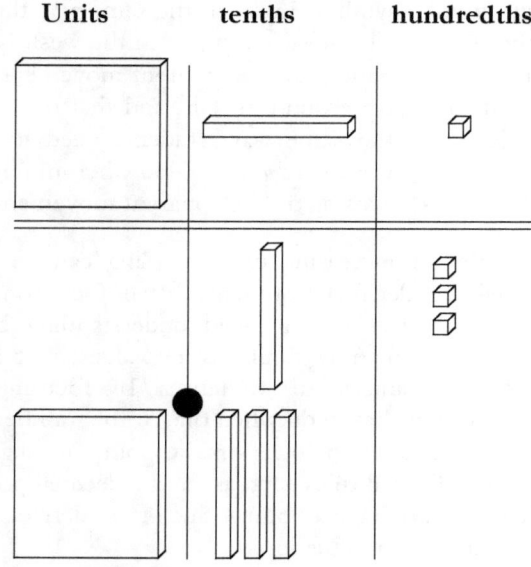

Figure 1. Blocks in row 1 are multiplied by 10 and the result represented in row 2

units		tenths	hundredths
	●	1	3
1	●	3	

Figure 2. The result of multiplying .13 by 10 is written in the second row

These teachers realized for the first time that the effect of multiplying a number by 10 could be understood in terms of looking at the value of the digits before and after the multiplication by 10, rather than looking at the point. They saw that students would have a deeper understanding of multiplication of a number by 10 if they would describe it in terms of moving each digit to the next place value column, rather than talking in terms of moving the decimal point.

"Moving the decimal point" is just a convenient way of talking about and operating with written symbols in much the same way that it is convenient to say that the sun rises in the east and sets in the west. Neither is sensible, students need to understand that the apparent movement of the sun is due to a rotation of the Earth around its axis, and that the sun is not orbiting around the Earth. In the same way, students need to understand what happens to the digits in terms of place value when multiplying or dividing by 10, rather than just focusing on the apparent movement of the point.

Another example for increasing understanding can be given for adding numbers containing decimals. Students often focus on aligning decimal points. They would get an improved understanding by describing the situation in terms of aligning similar place values, that is, adding tens to tens, units to units, and tenths to tenths. By focusing on place value, students would know what to do when one of the numbers does not have a decimal point. Students who focus on the points to align numbers, often need to rely on additional rules such as "add a decimal point to the right of a number that does not have a point." Such rules detract rather than add to their understanding of the situation.

There are other "moons" related to the decimal point. When dividing decimal numbers, students would have an enhanced understanding by thinking in terms of multiplying divisor and dividend by the same power of ten, and thus keeping the answer the same. Students who change a problem like $.25\overline{)1.25}$ into $25\overline{)125}$ can see why the answer is the same in both cases by thinking in the analogous situation of the multiplication of a fraction by 1 in the form of $\frac{100}{100}$, rather than focusing on the dragging of the decimal point the same number of places.

With decimals, students can change a multiplication like 1.2×1.4 into the corresponding multiplication 12×14 involving whole numbers. They can understand that with this change each factor has increased by a factor of ten, so that the answer, 168, is bigger by a factor of 100 than the answer corresponding to the original problem. Just focusing on counting decimal places to reinsert the decimal point is again to concentrate on the finger.

The decimal point is not the only "finger" when dealing with the "moon" of place value. Rules such as "add a zero" when multiplying a whole number times ten, also make students miss the central idea, that the value of each digit in the number has increased by a factor of ten and therefore moved to the next column. The zero holds a place for the units column, which is now empty because the digits have moved to columns where their value is ten times bigger.

What are other "fingers" and "moons" in our teaching of mathematics? Readers are invited to identify pointers that end up distracting students from understanding the important ideas, and to think about ways to help students concentrate on the ideas. Bruce Lee's advice to his student in *Enter the Dragon* was "Don't concentrate on the finger, or you will miss all the heavenly glory." In the same way, we need to guide our students not to concentrate just on the pointers or they will miss all the understanding.

References

Barnett, Carne, Donna Goldenstein, and Babette Jackson, eds. *Fractions, Decimals, Ratios, and Percents: Hard to Teach and Hard to Learn?* Portsmouth, NH: Heinemann, 1994.

Enter the dragon. Warner Home Video, 1998. (Originally issued in 1973)

Tanzania Proverb quotes. Retrieved February 2, 2006 from http://en.thinkexist.com/quotes/tanzania_proverb/

28 JELLY BEANS AND HAIRBANDS: AREAS OF TRIANGLES WITH THE SAME BASE AND SAME HEIGHT[28]

In the activities presented here, students in the sixth or seventh grades explore the area of a family of triangles all of which have the same base and the same height. A dynamical model made with an elastic hair band, four straws, three hair clips, and a shallow box or the lid of a rectangular cardboard box is used to represent several triangles in such family. The hair band is attached with two hair clips A and B (see Figure 1) to one wall of the box at two fixed places. The band is also attached to another hair clip P on the opposite wall, thus forming a triangle. The height of the triangle will be the width of the box. Two straws about 2 cm shorter than the height of the triangle are placed inside the hair band in the segments AP and BP to give some rigidity to the sides and maintain a triangular shape, and so that the hair band forms a wall to contain the jelly beans. Two rulers are taped next to each of the opposite walls. One ruler will be used to measure the base of the triangles; the second ruler will be used in another activity. To measure the area of the triangles, students will use non conventional units, namely collections of jelly beans. Students need to make sure they do not let go off the elastic hair band when changing the shapes, otherwise some jelly beans may get propelled.

---

[28] Yun, J. O. and Flores, A. (2007). Jelly beans and hair bands: Areas of triangles with the same base and same height. *Ohio Journal of School Mathematics*, No. 56, 33-37. Reprinted by permission.

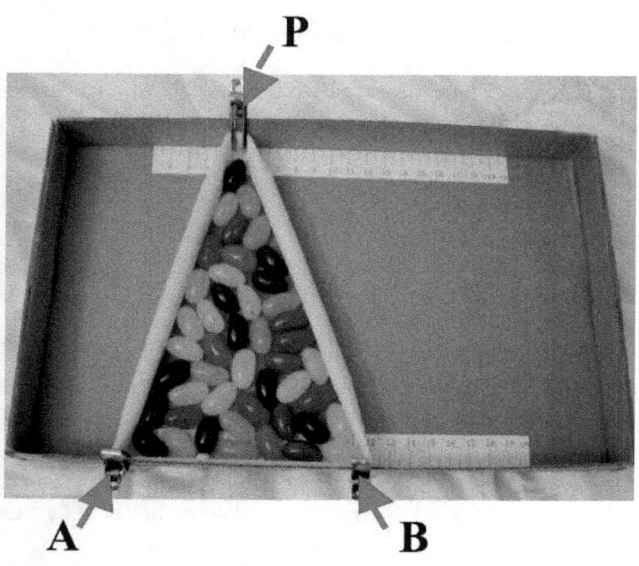

Figure 1. The area is covered by jelly beans.

Although jelly beans are three-dimensional objects, by using only one layer, we can use them to measure area. If all the jelly beans are on one layer, the more jelly beans we need to fill a two dimensional shape, the larger will be its area. Or in other words, by using one layer only of jelly beans we are essentially using the cross section of the jelly beans, which is two-dimensional, to cover the area. In this activity students do not have to count the number of jelly beans to compare areas; they use only the global amount. As they change the shape of the triangle, they will see whether all the jelly beans fit into the new shape or not. Because we are not counting jelly beans, it does not matter whether the jelly beans are all exactly the same size. Also, we are not comparing collections of jelly beans from one group of students to another. Each group will work with its own set, and determine only whether or not the total amount that fits in one shape fits into a second shape.

In later activities, students vary the base or the height of the triangles, and construct rectangles. Students will need an additional clip to form rectangles. Up to four students can work together using one box. The purposes of the lesson are that students

1) Verify visually that the areas of triangles do not change when the base and the height do not change in spite of the changing shapes of the triangles.

2) Understand what elements determine the area of a triangle.

3) Infer the formula of the area of triangle from the formula for the area of a rectangle.

Activities for students

Activity 1. Constant height, constant base.

1. Make a triangle inside the box with the given hair clips and a hair band. You can use the ruler to measure the base of the triangle.
2. Fill in the triangle with jelly beans (or other candies). Because the jelly beans are round, tiny spaces between them will not be covered, but make sure that the open spaces are quite smaller than the jelly beans. Make sure also that the jelly beans form just one layer, that is, jelly beans should not pile up on each other (Figure 1).
3. Slowly slide the hair clip P along the wall of the box while the other two clips A and B on the opposite wall remain fixed. Notice how the jelly beans rearrange themselves inside the new triangles (Figure 2). If needed, flatten out the jelly beans.

Do you need more jelly beans or fewer jelly beans when you move the point P?

Compare the base of the original triangle with the base of the new triangles. Compare the height of the original triangle with the height of the new triangles.

What can you say about the bases and the heights of these triangles?

P

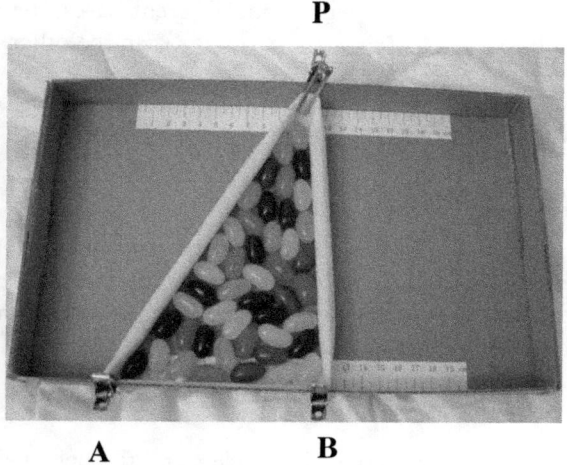

A   **B**

Figure 2. Another triangle with the same base and the same height.

Activity 2. Constant height, changing base

1. Now keep the hair clip A and hair clip P fixed. Move the hair clip B.

Do you need more jelly beans or fewer jelly beans when the length of the base of the triangle increases?

Do you need more jelly beans or fewer jelly beans when the length of the base of the triangle decreases?

Activity 3. Constant base, changing height

1. Move the hair clip P to somewhere inside of the box while A and B are fixed. You will need to keep clip P with your hand at a fixed position inside the box, rather than fixing it to the wall of the box.

Do you need more jelly beans or fewer jelly beans when the height of the triangle is reduced?

Discussion

Discuss the conditions by which the area of a triangle was changed or not, and fill in the blanks in the following sentences.

Triangles with same _____ and the same _____ have equal areas.

If two triangles have the same height, but different bases, the one with the larger base will have _____ than the other.

If two triangles have the same base, but different heights, the one with the larger height will have _____ than the other.

Activity 4. Area of rectangle

1. Make a **rectangle** whose base is half of the original triangle's base and whose height is the same as the height of that triangle. You can locate the fourth vertex by stretching the side of the hair band with your fingers or by using an extra hair clip.

Do you need the same amount of jelly beans or not? Is it always the case?

If we denote the base of the original triangle b and the height h, what would be the base and the height of the rectangle? What would be the area of the rectangle? Write a formula for the area of this rectangle in terms of the base and height of the original triangle. _____ What would be the area A of the original triangle? Write a formula for the area of the triangle in terms of its base and its height $A =$ _____

Advanced Question (adapted from Kang, Jung and Lee 2006, p. 78)

Peter and John own farms that have a border between them that is not straight (Figure 3). Peter and John want to make the border straight without changing the areas of their farms. Explain how they can solve their problem using some insight from activity 1.

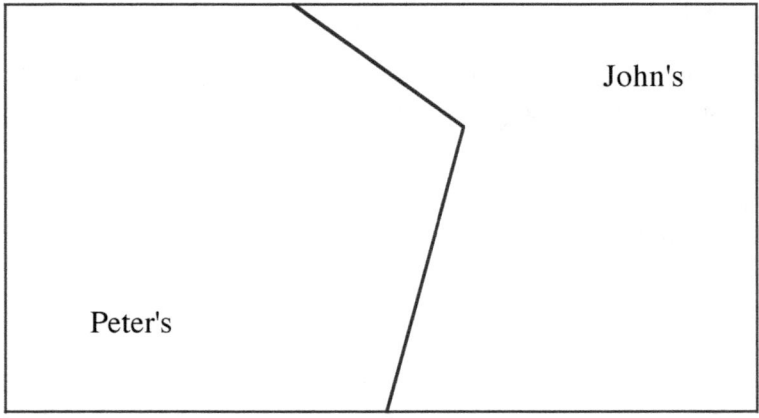

Figure 3. Two farms with a crooked border.

This problem is also appropriate for more advanced students. For 6th and 7th graders teachers may want to give students some pointers.

Hint for advanced problem. Through point P draw a parallel segment to AB. A parallelogram ABDC will be formed. Because parallelograms have constant width, any triangle with base AB and third vertex on segment CD will have the same height as triangle ABP, and therefore the same area.

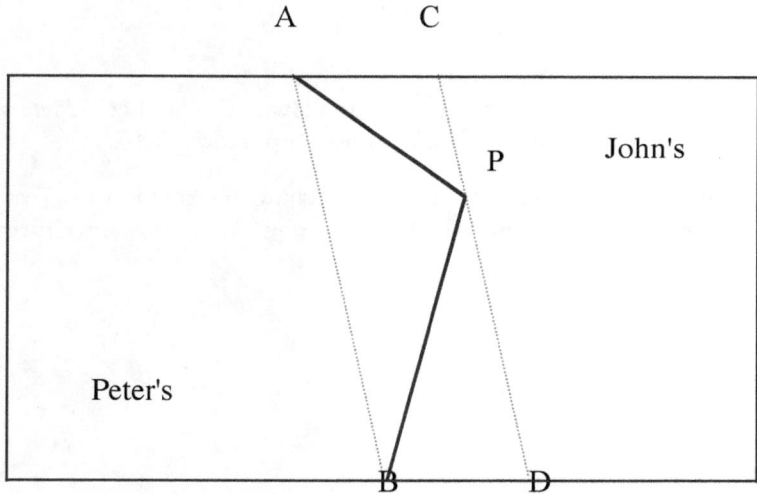

Figure 4. A hint for students.

Final comments

Middle school students are pleasantly amazed to see how the jelly beans arrange themselves to fit the new triangle with the same base and same height and thus see that the area remains the same. Occasionally they will have to "flatten" the beans to avoid some overlaps and make sure no large empty spaces remain, but in general it will become very clear if the jelly beans fit into the new shape or not. These activities are also very successful with prospective or in-service teachers (Yun, 2007). Even though they may know that the area of the triangle depends only on its base and height, they still enjoy seeing how the jelly beans fill the shapes that have the same area. Of course, as any concrete model that represents abstract mathematical ideas, using jelly beans to measure area has limitations. If the angle at P becomes too small, jelly beans will not fit between the sides of the triangle close to that vertex, so a significant part of area of the triangle will not be covered by jelly beans.

The first three activities with triangles will show that the area of a triangle depends only on its base and height, but they do not lead to an explicit formula of how base and height can be used to determine the area. The fourth activity allows students who know the formula for a rectangle to use it to compute the area of a triangle.

References

Kang, Ok-Ki, Soon-Young Jung, and Hwan-Cheol Lee. *Mathematics Textbook 8-na.* Seoul, Korea: Doosan Corporation, 2007.

Yun, Jeong Oak. The area of a triangle. Teaching presentation for Teaching Geometry 5-8. Unpublished manuscript. Arizona State University, 2007.

29 ALGEBRA FOR A PRINCESS[29]

Seldom in the history of mathematics we have a book of arithmetic and algebra named after a beautiful young girl. Bhaskara the learned (also known as Bhaskaracharya), named his book *Lilavati* after his daughter. Her name means beautiful. Bhaskara was a mathematician and astronomer who lived in India between the years 1114 and 1185 of the western count. Several of the problems in the book are explicitly addressed to her.

> **16.** Beautiful and dear Lilavati, whose eyes are like a fawn's! tell me what are the numbers resulting from135 [multiplied by] 12? .... Tell me, auspicious woman, what is the quotient of the product divided by the same multiplier? (Bháscara, 1971, p. 6)

Other problems let us know more about this charming girl. She is described as "pretty girl with tremulous eyes" (example 49), and as a lovely woman, intelligent, and skillful calculator (see examples 13, 37, 58, 68 at the end of the article). One cannot help but wonder. Who was this pretty girl? Why does she have tremulous eyes? Why is she asked to solve problems like these?

We also learn a little about the life around Lilavati. Examples are often set in rich, splendorous, and romantic settings

> **100.** Four jewelers, possessing respectively eight rubies, ten sapphires, a hundred pearls, and five diamonds, present, each from his own stock, one apiece to the rest in token of regard and gratification at meeting: and they thus become owners of stock of

[29] Flores, A. (2007). Algebra for a princess. *The Centroid, 33*(1), 5- 8. Reprinted by permission.

precisely equal value. Tell me, severally, friend, what were the prices of their gems respectively? (Bháscara, 1971, p. 45)

114. In a pleasant, spacious and elegant edifice, with eight doors, constructed by a skilful architect, as a palace for the lord of the land, tell me the permutations of apertures taken one, two three, &c. (Bháscara, 1971, p. 50)

This example is from one of the commentaries

The third part of a necklace of pearls, broken in an amorous struggle, fell to the ground: its fifth part rested on the couch; the sixth part was saved by the [young woman]; and the tenth part was taken up by her lover: six pearls remained strung. Say, of how many pearls the necklace was composed (Bháscara, 1971, p. 25, footnote 5)

However, there are also reminders that in those days life could be very harsh for some people, like the slaves. The inverted rule of three terms is used to figure out the value of living beings, when their value is regulated by their age, depending on the number of service they may provide in the future. So, the older a person, the less her value, as in the following example, where a slave age 20 is worth less than a slave age 16.

76. If a female slave sixteen years of age, brings 32 [nishcas], what will one aged twenty cost? (Bháscara, 1971, p. 34)

About the mathematics in *Lilavati*

One is also intrigued by the mathematics itself. Let us look at a multi-step problem.

49. Pretty girl with tremulous eyes, if thou know the correct method of inversion, tell me, what is the number, which multiplied by three, and added to three quarters of the [product], and divided by seven, and reduced by subtraction of a third part of the quotient, and then multiplied into itself, and having fifty-two subtracted from the product, and the square root of the remainder extracted, and eight added, and the sum divided by ten, yields two? (Bháscara, 1971, p. 21-22)

How can one solve such problem? What exactly is one asked to do in each step? How can we relate the methods used in India in the 12th century to our own modern symbolic methods? The method used by Bhaskara to solve this problem is by inverting the operations involved. Bhaskara gives the rule of inversion as follows. The first part of the explanation is pretty straightforward.

> To investigate a quantity, one being given, make the divisor a multiplicator: and the multiplier, a divisor; the square, a root; and the root, a square; turn the negative into positive; and the positive into negative. (Bháscara, 1971, p. 21)

Lilavati was a book that had a great influence in Indian mathematics. Several commentators added notes and explanations, because the original was not always clear for everybody. Here is part of the explanation of the rule of inversion which may not be as clear for modern students.

> If a quantity was to be increased or diminished by its own proportionate part, let the denominator, being increased or diminished by its numerator, become the denominator, and the numerator remain unchanged; and then proceed with the other operations of inversion, as before directed. (Bháscara, 1971, p. 21)

Here are some clarifications that may help understand these instructions of the inversion method. We will describe them also using modern mathematical notation. For example, a number added to three quarters of the number is the same as the number times seven quarters, $p + \frac{3}{4}p = \frac{7}{4}p$.

We can see that the denominator of the second fraction is the same as the denominator of the first. And 7, the numerator of the second fraction is obtained by adding the numerator and denominator of the first fraction. A number reduced by a third part of the number is the same as multiplying the number by two thirds, $r - \frac{1}{3}r = \frac{2}{3}r$. Here again, the denominators of the two fractions are the same, and the numerator of the second fraction is obtained by subtracting the numerator of the first fraction from its denominator. We can tabulate the operations of the multistep problem using letters for the unknown numbers at different steps of the process.

Initial number	N
multiplied by 3	$3n = p$
added to three quarters of the product	$p + \frac{3}{4}p = \frac{7}{4}p = q$
divided by seven	$q / 7 = r$

reduced by subtraction of third part of the quotient	$r - \frac{1}{3}r = \frac{2}{3}r = s$
multiplied into itself	$s \times s = t$
having 52 subtracted from the product	$t - 52 = u$
square root of the remainder extracted	$\sqrt{u} = v$
eight added	$v + 8 = w$
the sum divided by ten yields two	$w / 10 = 2$

To solve the problem we need to use the inverse operation for each step. The chain of inverted operations is thus

$$\times 10 \qquad -8 \qquad \text{square} \qquad +52 \qquad \text{sq root} \qquad \times 3/2 \qquad \times 7 \qquad \div 7/4 \qquad \div 3$$
$$2 \rightarrow 20 \rightarrow 12 \rightarrow 144 \rightarrow 196 \rightarrow 14 \rightarrow 21 \rightarrow 147 \rightarrow 84 \rightarrow 28$$

The method described by Bhaskara to solve this problem is very close to the process known as backtracking. "Backtracking is a process of finding an unknown number, n, by working backwards from the result, using students' natural problem solving skills. It breaks complex algebraic equations down into stages" (Lovitt and Clarke, 1988).

In the example above, instead of using several different letters we can also describe the process with only one letter to represent the unknown number,

$$\frac{\sqrt{\left(\left(\frac{2}{3} \times \frac{\frac{7}{4} \times 3n}{7} \right)^2 - 52 \right) + 8}}{10} = 2.$$ A closely related and very effective way for

beginners to deal with such complex equations is to use the cover up method (Whitman 1982; Kieran and Chalouh 1999). If in the above equation we cover the numerator of the left side expression with a box, it

becomes a less intimidating equation, $\dfrac{\square}{10} = 2$. What is inside the box is

thus equal to 20. Thus $\sqrt{\left(\left(\frac{2}{3} \times \frac{\frac{7}{4} \times 3n}{7} \right)^2 - 52 \right) + 8} = 20.$ Covering the

square root, the equation becomes $\square + 8 = 20$. By continuing using the covering method we can reverse the steps one by one until we obtain $3n = 84$, and finally, $n = 28$.

As can be seen from the sample of problems given, *Lilavati* has a wealth of different kinds of arithmetic and algebraic problems. Teachers can choose from among the problems. Students can state them into their own mathematical notation and solve them using their own methods. Students can also represent the relationship between the quantities in the problems using modern mathematical notation. Example 58 (below) can be stated as $a - b = 8$, $a^2 - b^2 = 400$. Example 68 (below) can be represented by the equation $n - \sqrt{\dfrac{n}{2}} - \dfrac{8}{9} = 2$. Example 76 (above) can be represented by an inverse proportion, $\dfrac{16}{20} = \dfrac{x}{32}$.

Bhaskara was also a poet. *Lilavati* was written in verses. He was obviously proud of his daughter and also of his book. In the dedication of the book, the author describes his process of computation, or arithmetic, as "delightful by its elegance, perspicuous with words concise, soft and correct, and pleasing to the learned" (p. 1). He ends the book with this remark.

> Joy and happiness is indeed ever increasing in this world for those who have *Lilavati* clasped to their throats, decorated as the members are with neat reduction of fractions, multiplication and involution, pure and perfect as are the solutions, and tasteful as is the speech which is exemplified. (Bháscara, 1971, p. 277)

It is nice to see an author who deals with this topic in such a pleasurable way.

Bhaskara and Lilavati are not only historic figures, but also legendary. In those times astronomers used to be also astrologers. Here is the story according to Joseph (1992) about the origin of the book's name. Slightly different versions of the legend exist. From casting Lilavati's horoscope, Bhaskara

> discovered that the auspicious time for her wedding would be a particular hour on a certain day. He placed a cup with a small hole at the bottom in a vessel filled with water, arranged so that the cup would sink at the beginning of the propitious hour. When everything was ready and the cup was placed in the vessel, Lilavati suddenly out of curiosity bent over the vessel and a pearl from her dress fell into the cup and blocked the hole in it. The lucky hour

passed without the cup sinking. Bhaskaracharya believed that the way to console his dejected daughter, who now would never get married, was to write her a manual of mathematics! (Joseph 19992 p. 269)

Additional Examples

13. Dear intelligent Lilavati, if thou be skilled in the addition and subtraction, tell me the sum of 2, 5, 32, 193, 18, 10, and 100 added together; and the remainder when their sum is subtracted from 10 000.

37. Tell me, dear woman, quickly, how much a fifth, a quarter, a third, a half, and a sixth, make when added together. Say instantly what is the residue of three, subtracting those fractions?

58. Tell me quickly, skillful calculator, what number are they, of which the difference is 8, and the difference of squares 400.

68. The square root of half the number of a swarm of bees is gone to a shrub of jasmine; and so are eight-ninths of the whole swarm: a female is buzzing to one remaining male that is humming within a lotus, in which he is confined, haven been allured to it by its fragrance at night. Say lovely woman, the number of bees.

The examples given in this article follow mainly the wording and numeration of Colebrooke' translation. Additional examples are available on-line with slightly different numeration and wording (History of Mathematics at Brown).

References

Bháscara. (1971). Arithmetic (Lílávatí). In H. T. Colebrooke (Ed.), *Algebra, with arithmetic and mensuration from the sanscrit of Brahmegupta and Bháscara* (pp. 1-127). Ann Arbor, MI: University Microfilms.

History of Mathematics at Brown. (n.d.) "*Lilavati*: The eight arithmetic operations." On line] Available http://www.brown.edu/Departments/History_Mathematics/lila/lila vati_eightops.html

History of Mathematics at Brown. (n. d.). "*Lilavati*: Standard computational techniques." [On line] Available http://www.brown.edu/Departments/History_Mathematics/lila/lila

vati_stdtechs.html October 17, 2006.

Joseph, G. G. (1992). *The crest of the peacock. Non-European roots of mathematics.* London: Penguin Books.

Kieran, C., & Chalouh, L. (1999). Prealgebra: The transition from arithmetic to algebra. In B. Moses (Ed.), *Algebraic thinking Grades K-12* (pp. 59-70). Reston, VA: National Council of Teachers of Mathematics.

Lovitt, C., & Clarke, D. (1988). *The Mathematics Curriculum and Teaching Program. Activity Bank* (Vol. 1). Canberra, Australia: Curriculum Development Centre.

Whitman, B. S. (1982). Intuitive equation-solving skills. In L. Silvey (Ed.), *Mathematics for the middle grades (5-9)* (pp. 199-204). Reston, VA: National Council of Teachers of Mathematics.

ABOUT THE AUTHOR

Alfinio Flores is professor of Mathematics Education at the University of Delaware. He has Mathematics degrees from the National University in Mexico and in Mathematics Education from Ohio State University. He teaches both mathematics methods and content courses *con ganas*. He uses technology, multiple approaches and concrete materials to develop conceptual understanding of mathematical ideas. He has published over 120 articles about the teaching and learning of Mathematics. He has conducted workshops for students ranging from Kindergarten to College, and taught courses for pre-service and in-service teachers in 32 states in two countries.